Biology of Reproduction

TERTIARY LEVEL BIOLOGY

A series covering selected areas of biology at advanced undergraduate level. While designed specifically for course options at this level within Universities and Polytechnics, the series will be of great value to specialists and research workers in other fields who require a knowledge of the essentials of a subject.

Titles in the series:

Biological Membranes	Harrison and Lunt
Water and Plants	Meidner and Sheriff
Comparative Immunobiology	Manning and Turner
Methods in Experimental Biology	Ralph
Experimentation in Biology	Ridgman
Visceral Muscle	Huddart and Hunt
An Introduction of Biological Rhythms	Saunders
Biology of Nematodes	Croll and Matthews
Biology of Ageing	Lamb
Biology of Reproduction	Hogarth
Introduction to Marine Science	Meadows and Campbell
Biology of Fresh Waters	Maitland

Biology of Reproduction

PETER J. HOGARTH, B.Sc., D. Phil.

Lecturer in Biology
University of York

A HALSTED PRESS BOOK

John Wiley and Sons

New York -Toronto

Blackie & Son Limited
Bishopbriggs
Glasgow G64 2NZ

450 Edgware Road
London W2 1EG

Published in the U.S.A. and Canada by
Halsted Press,
a Division of John Wiley and Sons Inc.,
New York

Library of Congress Cataloging in Publication Data
Hogarth, Peter J.
Biology of reproduction

(Tertiary level biology)
"A Halsted Press book"
Bibliography: p.
Includes index.
1. Reproduction. I. Title.
QP251.73   596'.01'6   78–4620
ISBN 0–470–26314–8

Printed in Great Britain by
Robert MacLehose and Company Limited

Preface

RECENT YEARS HAVE SEEN GREAT PROGRESS IN OUR UNDERSTANDING OF THE processes involved in reproduction. This book is an attempt to present, concisely but reasonably comprehensively, the current state of knowledge in an exciting and rapidly advancing subject.

In writing it, I have tried to avoid two major temptations. On the one hand is the lure of the laboratory rat. So much is known about this species (and about the handful of other species which contribute disproportionately to the research literature) that it is very easy to fall in to the trap of discussing the minutiae of rat reproduction to the extent that other species seem merely incidental. On the other hand, overemphasizing the diversity of reproduction among the mammals can result in the sacrifice of depth to breadth. I hope I have managed to follow a coherent path between these extremes.

I have naturally drawn heavily on my own experience of teaching reproductive physiology to successive classes of final-year undergraduates at the University of York. Primarily, therefore, the book is aimed at advanced university or polytechnic students in biology, zoology, or a related subject. I hope it may also serve to introduce the complexity and fascination of reproductive biology to those working in other fields.

I am grateful to the numerous undergraduates who, by their comments and suggestions about my lectures, have unwittingly helped me to write this book. I should also like to thank Dr Linda Partridge, Dr Henry Leese, and my wife, Dr Sylvia Hogarth, for reading various chapters and commenting on them, Paul Wheater for generously supplying me with the material for figure 2.3, Mr R. Hunter for his skilled artwork, and the Publishers for their help and encouragement.

P. J. H.
Sept. 1977

Contents

CHAPTER ONE

INTRODUCTION

TO UNDERSTAND FULLY THE BIOLOGY OF REPRODUCTION, IT IS NECESSARY TO take into account biochemical, cellular, physiological, behavioural and ecological considerations; sociology, psychology, and economics are also important in so far as they impinge directly on human reproduction. The breadth of the field is self-evident. It is well to consider also its inherent limitations.

Knowledge is limited by the nature of the technique available for acquiring it. Before modern biochemical methods took shape, for instance, the existence but not the chemical identities of hormones could be postulated. When appropriate biochemical techniques became available, not all hormones could be identified with equal ease. Gonadal steroids (sections 2.7 and 4.5) were readily identified, being simple in structure and present in relatively large quantities. The pituitary gonadotrophic hormones, being glycoprotein and occurring at lower concentrations than gonadal steroids, proved more recalcitrant. To accumulate a single milligram of hypothalamic releasing factor (section 2.7) for analysis, it is necessary to collect about 3 kg of pig hypothalamus!

Initially, hormones could be identified and measured only in terms of their biological effects. Bioassay techniques became highly sophisticated and precise, and are still used. A bioassay method, however, can measure hormone concentrations only relative to an arbitrarily defined standard. Precise determination of absolute, rather than relative, hormone concentrations could be achieved readily only after the development of techniques which exploited the specificity of antibodies or other naturally-occurring proteins with a specific binding affinity for hormones.

These techniques—particularly radioimmunoassay (RIA)—have in recent years yielded an enormous amount of otherwise unobtainable information (crucial to an understanding of the control of reproductive processes) about variations in hormone concentrations. One must not, however, become so mesmerized by precision as not to consider what is

1

actually being measured. It happens that in some circumstances (see, for instance, section 2.7) bioassay and radioimmunoassay estimates of hormone concentration do not agree. The biological hormone receptor and the antibody molecules of the RIA method may be sensitive to different parts of the hormone molecule, and changes may occur in either part of the molecule independently. Possibly a process of denaturation reduces the biological activity of a proportion of hormone molecules, leaving unchanged the parts of the molecule to which the antibody is directed, or vice versa. The biological property of the molecule matters more to the animal under study; the specificity and precision of RIA techniques is here less useful than the less accurate but more relevant bioassay.

Much the same could be said of other techniques. The advance of knowledge and understanding depends on technical advances, but these must have their limitations and must be critically assessed.

Other major limitations in reproductive biology are the restricted number of species which have been investigated in any depth, and the constraints on the way in which many of the popular species can be studied. Human reproduction has been studied in great detail, largely from a clinical point of view. The work has been largely observational, with relatively little specifically experimental work, and that often based necessarily on statistically inadequate samples. Good experimental design, with numerous replicates, is not always easy to achieve when research is (rightly) secondary to treatment. Certain categories of experiment are totally precluded. On the other hand, humans are large— which makes, for example, the collection of large numbers of serial blood samples for hormone assay more practicable than in a very small species; and, within limits, humans can usually be induced to cooperate in ways which other animals cannot. In short, the nature of the information which can be gathered is conditioned by the peculiarities of the species.

Domestic animals—cows, sheep and pigs in particular—have also been much used, largely because of the commercial importance attached to them. They, too, suffer from experimental drawbacks, such as expense. It is not surprising, then, that the overwhelming majority of experiments in reproductive biology are done on the standard laboratory animals such as mice and rats. These have the advantage of being small, cheap, and easy to breed. No species is representative of all mammals, and small rodents are in many respects more unrepresentative than most. To offset this, they do have the inestimable advantage of being traditional. It is always convenient to choose an animal about which much is already known. Scientists work on rats and mice largely because, for the last fifty years and more, scientists have been working on rats and mice.

A count of the animals used in a sample of more than 400 papers which appeared in randomly-chosen recent issues of the *Journal of Reproduction and Fertility*, showed that rats, mice, the common domestic animals and man together accounted for almost exactly 90%. The remaining mammalian species (more than 4200 of them) were altogether represented among a mere 10% of papers published. Obviously most species are virtually never looked at from the point of view of reproduction. This book attempts to introduce some of the nonconformist species and apparently anomalous forms of reproduction. This is not to introduce novelty for its own sake, but rather to exemplify the evolutionary point that, while the objectives of reproduction may be universal (one pair must leave on average at least two offspring), the means which have evolved to achieve the objective vary.

A comprehensive account of mammalian reproduction would include marsupials as well as eutherian mammals. Among the Eutheria, pregnancy in general determines the lifespan of the corpus luteum (section 6.1). With the marsupials, it is the other way about: the lifespan of the corpus luteum (with few exceptions) cannot be prolonged in the event of pregnancy. The marsupial gestation period is limited to the time for which a corpus luteum is active in the course of a normal reproductive cycle. What is, in effect, extension of the duration of fetal development can be achieved only outside the uterus, in a marsupial pouch. The contrasts between marsupial and eutherian reproduction emphasize the significance of the eutherian discovery of how to sustain a corpus luteum.

Consideration of the lower vertebrates adds to the evolutionary perspective. Viviparity is by no means the exclusive prerogative of mammals. Some members of every vertebrate Class, except Agnatha and the birds, give birth to live young rather than laying eggs, and it is quite clear that the viviparity has evolved independently numerous times. The problems posed by pregnancy—hormonal, nutritional, immunological—must therefore have been solved independently in each case.

It is not possible here to do more than select one or two examples. Some sharks and dogfish have a placenta which resembles in gross structure that of eutherian mammals, with an umbilical cord and a similarly intimate relationship with the lining of the uterus. In structure and function it is analogous to the mammalian equivalent, but in origin it derives from a different tissue structure. The lack of homology with a mammalian placenta indicates independent, albeit convergent, evolution. All viviparous vertebrates are faced with the problem of avoiding immune rejection of the fetus by the mother during pregnancy. In some viviparous Amphibia, this appears to be achieved (partly at least) by the induction of a state known as *specific immune enhancement*. Such a mechanism has also been implicated in survival of the mammalian fetus (section 9.5). There is no evolutionary continuity between viviparity in the amphibians and in mammals, so again the mechanism of avoiding fetal rejection has evolved independently in comparable circumstances. Similar solutions evolved to similar problems.

The diversity of reproduction, within the eutherian mammals and throughout the rest of the Animal Kingdom, should not obscure the underlying similarities.

Gamete dimorphism is universal in the vertebrates; small motile sperms are produced in prodigious numbers; ova are larger, fewer and immobile. Production of sperm in a sexually active male is continuous; mature ova are produced only intermittently. Consequently, female sexual behaviour is also variable, as is the physiological state of the female genital tract.

Regulation of reproductive processes is largely hormonal. In all vertebrates, gonads of either sex produce steroids which, acting as hormones, affect secondary sexual characteristics, the functioning of the reproductive tract, and sexual behaviour. Why gonadal hormones should, with very few exceptions, be steroids is not clear. Apart from the obvious point that steroids as a group have properties appropriate to the various functions they perform, it is possible that they originally had some role intrinsic to gametogenesis and that their hormonal actions external to the gonad evolved subsequently and secondarily to this.

In turn, production by the gonads of both gametes and hormones is under the control of pituitary glycoprotein hormones. Two major gonadotrophins, LH and FSH, are involved (see sections 2.7 and 4.3) with a third, prolactin, playing a rather ill-defined additional role. All three are found in mammals; in teleosts, a prolactin is present, but the functions of LH and FSH may be performed by a single molecule with many of the properties of both. If this is confirmed, it probably reflects the ancestral situation in the vertebrates, with a single gonadotrophin gradually diverging into two progressively more specialized entities.

Glycoprotein hormones are produced also by the placentae of many mammals. In the human, one of these, Human Placental Lactogen (HPL; see section 6.1.3) closely resembles pituitary Growth Hormone in its actions and structure, and the two probably share a common ancestry. The amino acid sequence of another placental glycoprotein hormone, Human

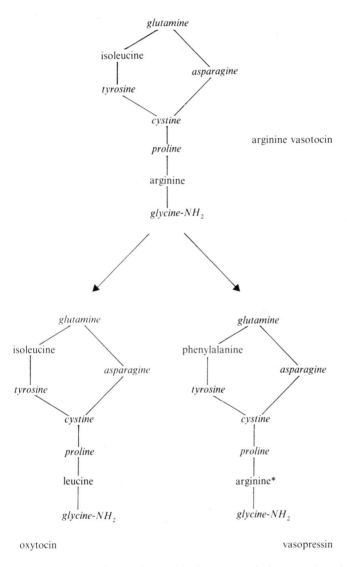

Figure 1.1 Relationship of the oligopeptide hormones of the posterior pituitary (neurohypophysis).

(*Arginine is replaced by lysine in pigs and hippopotamuses.)

Chorionic Gonadotrophin (HCG; see section 6.1.3) has been worked out in some detail. The molecule consists of two subunits. One, the α-subunit, closely resembles the α-subunit of human pituitary LH and that of pituitary Thyroid Stimulating Hormone, TSH. The other, the β-subunit, is also very like the β-subunit of pituitary LH, but has in addition about 30 amino acid residues not found in other β-subunits. Its homology with the β-subunit of TSH is much less. Carbohydrates associated with both HCG subunits are

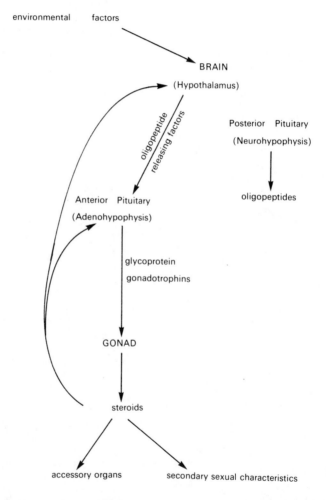

Figure 1.2   General scheme of the control of vertebrate reproduction.

different from those of the other related hormones. It is therefore possible, by studying the amino acid sequences of a number of related hormones, to postulate the phylogenetic relationship between them.

In turn, the pituitary hormones are controlled by oligopeptide releasing factors produced by neurosecretion in an adjacent part of the brain and conveyed to the anterior pituitary (adenohypophysis) by a portal blood system. The posterior pituitary (or neurohypophysis) or, in lower vertebrates, structures homologous with it, is in origin also neural tissue and likewise produces oligopeptides. Three oligopeptides have been identified in mammals (of which one, arginine vasotocin, appears to be produced by the pineal body). The three differ from one another only with respect to one or two amino acids; again, evolutionary relationships can be postulated (figure 1.1).

The brain-pituitary-gonadal axis (summarized in figure 1.2) appears to be universal throughout the vertebrates. Small amounts of releasing factors from the brain elicit the release of rather larger amounts of glycoprotein gonadotrophins and yet larger increases in gonadal steroids. The result is the amplification of small alterations in the activity of certain areas of the brain into large increases in circulating steroid hormones. This amplification makes the control of reproductive processes a delicate one. Steroids in turn act back in such a way as to determine the rate at which they themselves are produced. These two features—amplification and feedback control—seem to be common to all vertebrates; at least, all those that have so far been sufficiently investigated. Species which lay tens of thousands of eggs, those in which it is the male which bears the young, egg-laying mammals, marsupials and eutheria—all control and integrate their reproductive processes by means similar in outline if diversified in detail. Much of the fascination of the study of reproduction lies in the discovery of how this underlying system has adapted and developed to accommodate so many different patterns of reproduction.

CHAPTER TWO

MALE REPRODUCTION

2.1 Morphology of the male reproductive tract

The primary function of the testis is the production of spermatozoa capable of fertilization after deposition in the female genital tract. An ancillary function is the secretion of a fluid medium suitable for the suspension of the sperm, their carriage out of the testis into more distal parts of the male system, and their maintenance until ejaculation and, to some extent, after mating as well. To the spermatogenic and exocrine functions of the testis must be added a third, endocrine function: the sexual accessory organs, secondary sexual characteristics and, to some extent, sexual behaviour are under the control of hormones secreted by the testis.

The anatomy of the reproductive system of the male rat is shown in figure 2.1. The testes are suspended in a scrotum outside the abdominal cavity. A temperature lower than that of core body temperature appears to be essential for some stage of sperm production, although birds and many mammals (whales, elephants and shrews among them) reproduce successfully with abdominal testes. In species with scrotal testes, a heat-exchanging counter-current system serves to maintain a temperature differential with the abdomen, typically of about 2°C in man, and up to 10°C or more in some ruminants.

On the surface of each testis lies a long and convoluted tube known as the *epididymis* (plural: epididymides); distally, this enters the vas deferens. The two vasa join, receive the ureter from the bladder, and become the urethra, which runs through the penis to the exterior. In the region of the junction of the vasa with the urethra, a group of accessory glands add their secretions to the seminal fluid. These consist of the paired seminal vesicles, the prostate gland, and the paired bulbo-urethral or Cowper's glands. Dispersed among the urethral epithelium are small areas of secretory tissue, the urethral glands or Littré's glands.

Mammals vary considerably in the disposition and morphology of these

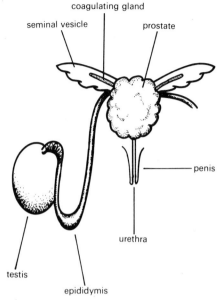

coagulating gland

seminal vesicle

prostate

penis

urethra

testis

epididymis

Figure 2.1 The reproductive tract of a male rodent

structures. Seminal vesicles are absent in cats and dogs, which have relatively enlarged prostates, while in the boar and bull the prostate is small but the seminal vesicles are well developed. In small rodents, a small coagulating gland lies on the surface of each seminal vesicle.

The function of the sexual accessory organs is discussed in section 2.6.

2.2 The testis and spermatogenesis

The testis consists of a fibrous capsule containing a mass of convoluted tubules of seminiferous epithelium which connect with a network of tubules, the rete testis. The rete in turn is connected to the epididymis by several excurrent ducts (figure 2.2). The arrangement has the effect of greatly amplifying the area of seminiferous epithelium and consequently the rate at which spermatozoa can be produced on it. For reasons which are not entirely clear (see section 2.3 and chapter 9), the mammalian testis produces sperm at a phenomenal rate, so that they can be ejaculated in seemingly extravagant numbers.

Figure 2.3 shows a transverse section through a number of testis tubules. Within each tubule lie two categories of cell: germ cells at various stages of

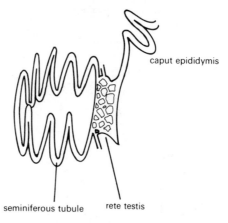

caput epididymis

seminiferous tubule   rete testis

Figure 2.2   Structure of a seminiferous tubule and its relation to the rete testis and epididymis.

development, and Sertoli cells, cytoplasmic processes of which are intimately associated with all stages of germ cell except the stem cells. The possible roles of Sertoli cells are discussed in section 2.4.

The male germ cells originate from gonocytes on the periphery of each tubule. These develop into spermatogonia, which may be subdivided into sequential stages of development as spermatogonia type A, intermediate, and type B. In some species, type A spermatogonia are further sub-divided, e.g., into A_1 to A_4 in the mouse. This phase of spermatogenesis is essentially one of multiplication, and the full sequence of spermatogonial development involves several mitotic cycles. However, not every type A spermatogonium will necessarily produce the maximum possible number of descendants; a proportion of type A spermatogonia return to the pool of stem cells, remain undifferentiated, and maintain the stock for future production of spermatozoa, while up to 90% fail to complete spermatogonial development and simply degenerate.

Halving of the chromosome complement is a necessary prerequisite to the doubling which will occur with fertilization and the fusion of the male and female pronuclei. The germ cells therefore next undergo meiosis. Type B spermatogonia divide to form primary spermatocytes. These divide with separation of the homologous paternal and maternal chromosomes to form secondary spermatocytes; the two sister chromatids of each chromosome then separate in the second division of the meiosis. The resultant cells are known as *spermatids*.

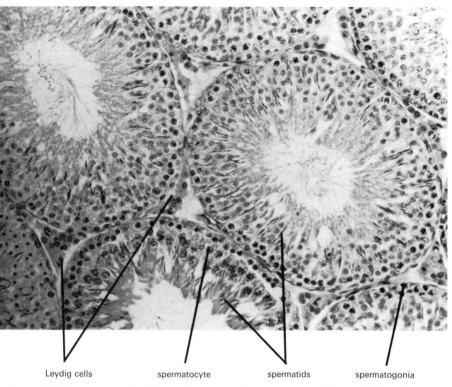

Leydig cells   spermatocyte   spermatids   spermatogonia

Figure 2.3 Transverse section of part of a mouse testis, showing seminiferous tubules and interstitial (Leydig) cells. (Photograph by P. Wheater).

The final phase of spermatogenesis in the testis is known as *spermiogenesis* (or spermateleosis), and consists of the progressive conversion, without division, of the spermatids into functional spermatozoa. Figure 2.4 shows the sequence of events in spermiogenesis. Notably, one of the two centrosomes of the spermatid extends to form the flagellum of the spermatozoon, the Golgi zone gives rise to the acrosome (whose function is described in detail in section 3.3), the nucleus becomes a condensed and frequently spindle-shaped structure, and finally the mature spermatozoon is released from its association with a Sertoli cell into the lumen of the tubule, much of its cytoplasm being retained by the Sertoli cell. The remainder, the "cytoplasmic droplet", is shed soon afterwards.

The three main cell types in spermatogenesis—spermatogonia, spermatocytes, and spermatids—therefore correspond to the three

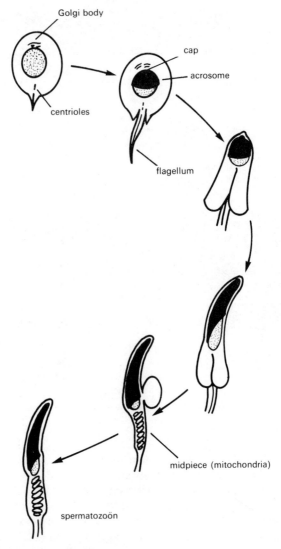

Figure 2.4   Spermiogenesis in the rat. (From Clermont, Y. (1967)).

necessary processes of multiplication (mitotically), meiosis (which also involves an increase in numbers), and maturation into functional spermatozoa.

A mature spermatozoon is surrounded throughout its life by a nutrient medium; little internal storage of metabolites is therefore necessary. Motility is supplied by the flagellum, and the energy for this by a small number of mitochondria. Penetration of the layers surrounding the ovum is achieved by acrosomal enzymes (section 3.3). Neither DNA, RNA nor protein synthesis are required, so it is not surprising that ribosomes and mRNA are absent, and that the haploid genome is condensed and inert, and remains so until after fertilization. The mature spermatozoon is a highly unusual cell adapted for a very brief period of activity after copulation.

2.3 Kinetics of spermatogenesis

Sperm are produced continuously throughout adult life, and consequently all stages of spermatogenesis are simultaneously present within the testis. (There are exceptions to this among seasonally-breeding mammals; perhaps among the most striking is the Rock Hyrax (*Procavia*) where the testes are normally regressed and periodically increase greatly in size for a brief period of sexual activity.

If a single seminiferous tubule is considered, this generalization is found to be untrue. Waves of spermatogenesis pass periodically along each tubule, so that within a particular region of tubule spermatogenesis appears to be initiated cyclically at regular intervals. The periodicity of these cycles has been determined for a number of species; in the mouse, rat (Wistar strain), ram and human, for instance, the intervals between successive initiations are 8·6, 13·3, 10·5 and 16 days respectively.

The synchrony of each successive cohort of germ cells is maintained throughout spermatogenesis; possibly the intracellular cytoplasmic bridges detectable by electron microscopy are important in coordination. Once a group of cells has started development, the time taken to form mature spermatozoa is constant, as is the duration of each successive stage of the process. For the four species mentioned above, the total times from initiation to the release of mature spermatozoa by the Sertoli cells are 34·5, 53·2, 49 and 74 days respectively. These figures demonstrate that the duration of spermatogenesis is several times as long as the cycle of initiation; before one cohort of germ cells attains maturity, several later generations will have started development and reached various stages of maturity. In any one section of a tubule, more than one stage of development is indeed present, with more advanced stages more centrally situated (figure 2.3).

In fact, because of the relative durations of the different stages of spermatogenesis, only a limited number of possible associations of germ cells is found. Figure 2.5 indicates the 14 possible cellular associations which occur in sequence in the seminiferous tubules of the rat. Conversely, of course, information about the incidence of different cellular associations in a species can be used to derive information about the periodicity of the cycle of the seminiferous epithelium or the duration of spermatogenesis.

The principal variant of this orderly process is found in the human testis, where no waves of initiation of spermatogenesis are detectable, and where specific arrays of germ-cell types are not found throughout an entire cross-section of a tubule. Instead, synchronized development of germ cells appears to take place in an irregular mosaic of small areas within the seminiferous epithelium. A single cross-section of a human tubule may include several such areas, and as a result a mixture of cell associations.

One of the most striking features of spermatogenesis is the enormous

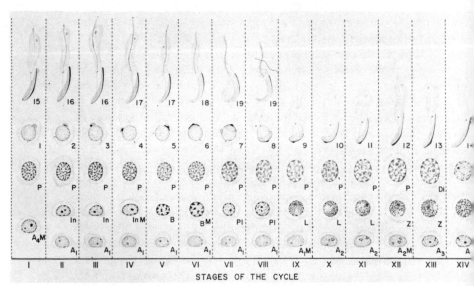

Figure 2.5 The fourteen cellular associations found in the seminiferous epithelium of the rat. Each column consists of the various cell types making a cellular association (identified by Roman numeral at base of column). The associations succeed one another cyclically in order from I to XIV. The complete sequence of changes undertaken by an individual germ cell can be seen by following the bottom row from left to right, then the next row above it, and so on. A, In, B refer to Type A, intermediate and Type B spermatogonia; R, L, Z, P, Di to primary spermatocytes at pre-leptotene, leptotene, zygotene, pachytene and diakinesis; II to secondary spermatocytes. The numbers 1 to 19 indicate distinguishable stages of spermiogenesis. (From Perey, B., Clermont, Y. and Leblond, C. P. (1961), *Amer. J. Anat.* **108**/1; 47–77 (fig. 2, p. 49).

number of spermatozoa produced. A human ejaculate, for example, may contain $3·5 \times 10^8$ sperm; of these, no more than 40 or so are ever found in the vicinity of the egg. Similarly profligate sperm production is found virtually throughout the animal kingdom. Why should so many sperm be produced?

One possible explanation is discussed in the next chapter. The time for which an ovum is available for fertilization is brief; in order to ensure the continuous presence in the oviduct of a quorum of fresh sperm it may be necessary for much larger numbers to be available in the "reservoir" regions of the female tract, such as the uterine cervix and the uterotubal junction (section 3.1).

Production of such vast numbers of sperm may also be necessary to offset the enormous losses which inevitably occur between sperm production and fertilization. If this mortality occurs simply because not all sperm have the opportunity to fertilize an egg, the argument would be a tautology. The fact that most of the sperm death actually occurs *before* the moment of fertilization indicates specific hazards overcome by only a small proportion of the sperm.

Copulation is not a sterile process; accordingly, the female genital tract is protected against infection by a variety of mechanisms. Among these are probably the low pH of the vaginal secretions and the cervical mucus (section 3.1), and the capacity of the female to mount a local immune response. At least some sperm must be able to survive these hazards. Evasion of the maternal immune response by sperm is discussed in section 9.2; it is clear, however, that whatever the precise details of the relations between sperm and female tract, the deposition of large numbers of sperm increases the likelihood that some among them will gain access to the egg.

Some losses of germ cells occur within the male genital tract, before ejaculation. Dead and degenerating cells are common even within the seminiferous tubules. In the rat, losses during spermatogenesis have been estimated at approximately 60% between early (type A_1) and intermediate spermatogonial stages, and 22% between late spermatogonial and early spermatid stages. An overwhelming majority of male germ cells fail to complete development. Other species show comparable sperm losses; large-scale cell death is a feature of spermatogenesis in all mammals so far studied, and in many other vertebrates and invertebrates.

Losses occur also after sperm have left the testis. In one series of experiments on a ram, for instance, the average daily loss of sperm in the urine accounted for virtually all the estimated daily production of spermatozoa.

Cohen has advanced a theory which might explain the apparent wastefulness of male reproduction at all stages of the life of the sperm. He suggests that during spermatogenesis, possibly stemming from the unusually high rate of cell division (particularly meiosis), a large number of "mistakes" of division occur. If the spermatozoa which resulted from a faulty meiosis (the vast majority of sperm, in Cohen's view) were permitted to fertilize ova, the resulting zygotes would not be viable. Clearly, it would be a more economic use of resources to prevent the formation of such unviable zygotes. The mortality of sperm which occurs in both male and female tracts might therefore be the result of selective filters for the elimination of genetically faulty sperm. The female genital tract has even been described by Cohen as "an 'obstacle race tract' rather than a device for promoting the union of egg and sperm. The Fallopian tube (= oviduct) in particular would appear to be an unnecessary complication." If only a tiny proportion of sperm are genetically "fit", in order to ensure that the absolute numbers of such fit sperm are sufficient for fertilization, production of vast numbers of sperm would be required.

If this interpretation is correct, then the greater the proportion of defective sperm produced, the greater the relative excess of sperm which will be required. The proportion of faulty sperm cannot be measured directly, but if it is true that most "errors" occur during the crossing-over at meiosis (which seems plausible), then the more frequently crossing-over occurs, the higher the proportion of defective sperm. It follows, therefore, that according to Cohen's theory, a correlation should exist between "gamete redundancy" (the numbers of sperm produced for every young born) and the chiasma frequency during gametogenesis. Such a correlation does exist (figure 2.6) (although of course there might be an alternative explanation).

Direct evidence for the selective nature of sperm death is sparse. In the rabbit, the sperm which reach the oviduct appear to be antigenically different from those which do not, and are also more fertile. This supports Cohen's view. On the other hand, selection can operate only on phenotypes, while the viability of the zygote once formed will be determined only by genotypic characteristics of the sperm. Selection of phenotypic properties of sperm would be irrelevant unless these were closely related to the genotype. In only a few cases has expression of the haploid genotype of spermatozoa (as opposed to the diploid genotype of premeiotic stages of spermatogenesis) been established with any degree of probability. Sperm in the ejaculate of many species are morphologically heterogeneous; in man, more than 10% of ejaculated sperm are abnormal

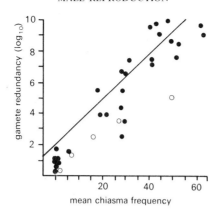

Figure 2.6   The relationship between "gamete redundancy" (the ratio of gametes produced to zygotes formed) and chiasma frequency per meiosis for a variety of species. Closed circles refer to sperm, open circles to eggs. (From Cohen, J. (1977)).

by gross morphological criteria, and presumably more stringent criteria would indicate a greater prevalence of abnormality. However, there is no reason to believe that selection by the female tract totally excludes these phenotypically abnormal sperm apart from the vicinity of the ovum. In mice, structurally abnormal sperm can not only reach the ova but also initiate development; following *in vitro* fertilization with such sperm, development proceeds as far as the two-cell stage with no indication of abnormality.

Evidence for a close relationship between the phenotype of sperm and their genotype is not good. Much of the excessive production of sperm is probably due to factors other than the need to select genetically "fit" sperm from among a genetically heterogeneous population produced by spermatogenesis.

2.4 Exocrine function of the testis

Spermatozoa leave the testis suspended, at a concentration of around 10^6/ml, in a fluid which is isotonic with blood plasma. This fluid is produced at the rate of $10 - 20 \ 1/g$ testis/h, or about 40 ml per day from a ram testis weighing 200 g. Although isotonic with blood plasma, it differs from it in ionic composition, and therefore cannot be simply a passive filtrate of plasma.

The precise location of secretion in the seminiferous tubules is not clear, although the Sertoli cells are the most likely source. Following treatment with the drug busulphan the tubules are virtually devoid of germ cells, but still contain fluid. However, there is reason to believe that backflow of fluid from the rete testis occurs, so the presence of fluid in the "Sertoli cell only" tubules is not conclusive.

Morphologically, the Sertoli cells resemble cells in other fluid-transporting tissues such as the gall bladder. The arrangement of Sertoli cells within a seminiferous tubule is shown in figure 2.7. This arrangement presents a number of interesting features. The presence of tight junctions separates the tubule contents into two compartments; spermatogonia lie in the basal compartment, while in the adluminal compartment are the spermatocytes and spermatids. The function of this separation is not known, but it may be that the diploid spermatogonia require a different ionic environment from the haploid spermatids and the spermatocytes undergoing meiosis, and that different environments are supplied by different regions of the Sertoli cell. It has also been suggested that the onset of meiosis might be controlled partly by the transition from the basal to the adluminal compartment.

The tight junction layer constitutes a major part of the "blood-testis

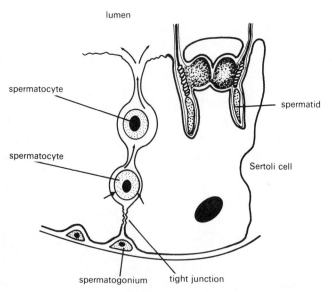

Figure 2.7   The relationship between Sertoli cells and developing germ cells (From Fawcett in Greep and Astwood (1975)).

barrier" which largely isolates developing germ cells, metabolically and immunologically, from the rest of the body. The dye acriflavine, injected into an adult male, penetrates most tissues readily, but is excluded from the lumen of the seminiferous tubules. In rats, an additional barrier is formed by the epithelial layer which surrounds the tubuies.

In a newborn rat or mouse, there is no blood-testis barrier, and acriflavine freely permeates the seminiferous tubules. By the age of 20 days, the barrier at the level of the tubules is complete; only after the occluding junctions appear, and acriflavine is excluded from the tubule, does fluid secretion commence and a lumen appear within the tubules. The tight junctions, as well as being an impediment to the passage of materials, may therefore also be necessary for the secretion of fluid into the lumen. According to one model of secretion, ions are actively secreted by Sertoli cells on the luminal side of the junction layer. Water follows passively, by osmosis. Positive hydrostatic pressure develops; because of the presence of the tight junctions, fluid flows only towards the lumen (figure 2.7). The nature of the secreted ions is not known, but since the rate of fluid secretion is unimpaired by ouabain (which eliminates Na^+ transport), sodium ions are probably not involved. Other possible ions are potassium and bicarbonate.

Micropuncture techniques have made it possible to collect for analysis fluid from the lumen of the seminiferous tubules. In addition to sampling this "free-flow" tubule fluid, small amounts of freshly-secreted "primary" fluid can also be obtained. This is done by first injecting a column of oil into the lumen; as fluid is secreted into the tubule, it collects in droplets which break the column of oil. The droplets can then be withdrawn and contain tubule fluid free from any mixture with liquid flowing back from the rete testis.

The composition of blood plasma is compared with that of the testis fluids, with respect to the major ions, in figure 2.8. Primary tubule fluid is relatively rich in K^+ and HCO_3^- ions, at the expense of Na^+ and Cl^-. Protein is virtually undetectable. Fluid sampled from the rete is closer in composition to blood plasma, and free-flow tubule fluid intermediate between primary tubule and rete fluid. All fluids are isotonic with plasma.

The most probable interpretation of these findings is that two different fluids are secreted, one in the seminiferous tubules and the other in the rete testis, and that backflow of the rete fluid into the tubules causes mixing and the presence there of a fluid intermediate in composition between the two primary secretions. If this is correct, the relative proportions of the ions in mixed fluid indicates that the rete contribution to the total volume is about

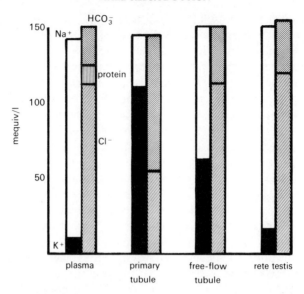

Figure 2.8 Ionic composition of fluids of the male genital tract, compared with that of blood plasma. Primary tubule fluid is that actually secreted in the seminiferous tubules; free-flow tubule fluid results from mixture of rete fluid with primary tubule fluid, and is the medium to which the germ cells will generally be exposed. (After Setchell and Waites in Perry (1971)).

ten times that of the tubules. The loop structure of the seminiferous tubules (figure 2.2) is suitable for the recirculation of rete fluid, and valve-like structures at the ends of the tubules may serve to prevent the recirculation of sperm.

Fluid sampled from the rete shows a number of other interesting differences from blood plasma. No glucose or fructose is present, but there are very high concentrations of another carbohydrate, myoinositol ($C_6H_{12}O_6$), about 100 times more than in plasma. Most amino acids are present in amounts smaller than the plasma concentration, but some at markedly higher levels: glycine ($\times$ 16 plasma level), proline ($\times$ 7·7), lysine ($\times$ 6·5), alanine and aspartic acid ($\times$ 4·5) in the rat. In the ram and bull, similarly elevated levels occur, but with glutamic acid replacing proline and lysine.

Not all the components of the tubule and rete fluids have functions reliably assigned to them. One function of the fluids is as a medium in which the sperm can survive their passage out of the testis, and such a medium would have to be carefully controlled with respect to osmotic concentration and pH. The presence of certain of the components will be related to this, for example HCO_3^- is involved in pH buffering of the blood, and might similarly buffer seminal pH. The compounds whose concentrations differ most obviously from plasma probably have other functions.

In vitro, boar sperm have been shown to oxidize a variety of amino acids, including glycine, proline, and alanine, which are present in tubule fluid. It is difficult to judge how important amino acids are for the respiration of spermatozoa in natural circumstances. Glutamic acid, for instance, is not used to a significant extent by sperm *in vivo* (and is in fact actually synthesized from glucose by sperm). Aspartic acid, glycine and glutamic acid are precursors of purine and pyrimidine bases, and their presence in the testis fluid may therefore be related to active DNA systhesis within the tubules, rather than to oxidative metabolism.

Inositol, another low-molecular-weight carbon compound which might seem suitable as a substrate, is also scarcely used. The reason for its presence in such high levels is unknown, although it is a constituent of certain phospholipids, and sperm depend heavily on endogenous phospholipid reserves as a supply of oxidizable substrate.

2.5 The epididymis

The epididymis is a highly convoluted tube connecting the testis to the vas deferens (figure 2.1). Along it, sperm in their suspension fluid are carried by peristaltic contractions; in the rat these occur with a frequency of one every 6–10 seconds. Cilia are present in the epididymal epithelium, but are non-motile and may serve either to prevent any backflow of sperm, or to mass the sperm together in the centre of the lumen.

The enormous length of the epididymis (7 metres in man, 86 metres in the stallion) means that transit of sperm takes a considerable time. This has been estimated for a number of species. Surprisingly, over a wide range of epididymal lengths (a bull epididymis, for example, being 20 times as long as that of a rabbit) the average transit times of sperm vary only from 8–14 days. Frequent ejaculation reduces this time by only 10–20%.

Sperm are produced continuously, but ejaculated only at irregular intervals. Accumulation and storage of sperm occurs in the cauda epididymis; from here they are expelled into the vas deferens during copulation. In the rabbit and guinea pig, sperm can remain viable for several weeks in the cauda, and hibernating male bats frequently retain sperm for several months. Probably the normal storage time is rather short. Species with abdominal testes (see p. 8) generally have a cauda displaced from the surface of the testis and lying next to the abdominal wall, which may indicate that it is storage of sperm, and not their production, which requires a temperature lower than that of the body.

In a celibate male the sperm which are not ejaculated are lost either in the urine or by phagocytosis and resorption in the epididymis or vas. In the ram, the former means of disposal accounts for virtually all of the production of sperm by the testis. The importance of phagocytosis is not clear; there is evidence that in some species it does occur, and even that abnormal sperm are preferentially removed. Even in sexually active animals, a significant proportion of sperm are lost by means other than copulation.

The composition of the seminal fluid alters during passage of the epididymis. Much water is withdrawn; in the ram, 40 ml enters from the testis every day but (even in a sexually active animal) only 0·4 ml enters the vas. The remaining water is resorbed in the epididymis. Ligation of the duct at the junction between testis and epididymis causes fluid accumulation in the testis, which does not occur if the ligature is placed at the junction between caput and corpus. Supported by cytological evidence of pino-cytosis, this indicates that the bulk of the water is resorbed in the caput. Withdrawal of sodium ions also alters the composition of the fluid. Relative to sodium, the concentration of potassium increases, partly as a result of the withdrawal of Na^+, but also partly because of the active secretion of K^+.

The osmotic pressure of the fluid in the cauda is higher than that estimated from the known concentrations of ions in the fluid, which indicates the secretion of further osmotically active substances. Three specific substances are now known to be added to the seminal fluid as it passes through the epididymis: glycerylphosphorylcholine (GPC), carnitine and glycoprotein.

The first of these may be present in relatively large amounts. In the epididymis of the boar, concentrations greater than 3g/100 ml have been measured. Part of this may be a product of phospholipid metabolism by the sperm, but the bulk is probably secreted by the epididymal epithelium. GPC cannot be used as a substrate by sperm within the epididymis. Enzymes suitable for GPC breakdown are, however, present in the female genital tracts of a number of species, which suggests that GPC may become available to sperm only after copulation. However, the presence of such enzymes has not been shown in all species, and the species in which they do occur are not invariably those in which the male has large amounts of seminal GPC (section 3.2). Alternative functions of GPC are in maintaining the osmotic pressure of seminal fluid, and in stabilizing spermatozoal membranes.

Carnitine is also present in large amounts. It is involved in fatty-acid metabolism (specifically, in Coenzyme A metabolism) and since sperm depend heavily on metabolizing phospholipids as an energy source, it might have an important role, but this has not yet been clarified. The function of the secreted glycoprotein is also unknown, although it may act as a lubricant for the sperm.

Ejaculated sperm can fertilize eggs, sperm collected directly from the testis cannot. The final maturation of spermatozoa, and their acquisition of full fertility, is therefore another process which occurs within the epididymis. Despite a great deal of research into such an obviously fundamental phenomenon, its nature is not yet clear.

As maturation proceeds, concomitant changes in structure and metabolism also occur. The last residue of cytoplasm, the "cytoplasmic droplet" is shed; in some species, the acrosome alters in appearance. Changes have been detected in membrance characteristics, specific gravity and protein content of sperm. Increased formation of disulphide cross-links stabilizes the nucleus and mechanically strengthens other parts of the spermatozoan. The capacity for full motility develops. In rabbits, testicular sperm can only vibrate their tails weakly; sperm from the caput epididymis can swim, but only in circles; vigorous directional swimming is achieved only when the sperm have reached the cauda.

Complex metabolic changes also occur. For example, testicular sperm synthesize amino acids and inositol from glucose; mature sperm do not. Testicular sperm depend more on oxidative metabolism when glucose is present, while mature sperm are predominantly glycolytic in their assimiliation of glucose. Although most information comes from the culture of sperm *in vitro* in conditions dissimilar from those within the epididymis (where, for example, only negligible amounts of glucose are present), it is clear that in many respects the metabolic capabilities of sperm alter with maturity.

Very little is known about the role of the epididymis in promoting sperm maturation. Rabbit sperm detained artificially in the rete testis, or in any part of the proximal epididymis, will eventually attain full motility. All that the epididymis supplies appears to be a congenial environment in which the sperm can pass sufficient time for their full motility to develop. The impressive length of the mammalian epididymis may be needed simply to give the sperm time to mature.

However, rabbit sperm held in the rete or proximal epididymis until they have developed full motility still lack the ability to fertilize eggs. Unlike motility, fertility of sperm appears to require more than time to be acquired. Some specific features of the environment in the more distal portions of the epididymis are necessary, but nothing is known of its nature.

It is apparent that, although much is known of the changes which take place in the sperm and seminal fluid within the epididymis, little is known of how these changes are effected, and of how they relate to each other.

2.6 Accessory glands

Most of the volume of semen (95–98% in the majority of species studied) comes not from the testis and epididymis but from the secretions of the accessory glands, and is added to the epididymal contribution during ejaculation. In animals such as boars and men in which ejaculation lasts for some time, it is possible to collect the ejaculate as a series of fractions, and therefore to determine the order in which the different components are contributed. Firstly, small volume of largely sperm-free fluid emerges. This is probably supplied mainly by the urethral glands. Following it is a fraction rich in sperm with contributions both from seminal vesicles and prostate; finally, a sperm-free ejaculate is added, predominantly of vesicular origin. A species such as the bull normally ejaculates instantaneously, but it is still possible in experimental conditions to demonstrate the sequential addition of the seminal components. Figure 2.9 shows the composition of five successive fractions obtained from a bull electrically stimulated to ejaculate.

Fructose and citric acid are generally among the major constituents of accessory gland secretion (Table 2.1) but the principal source of these varies from species to species. In most species, fructose is chiefly in the seminal vesicle secretion, but in rats it is produced in the coagulating glands, while citric acid is contributed by the seminal vesicles of boars and bulls, but by the prostate in man.

Fructose is presumably present as a nutrient for sperm after copulation. It is at highest concentrations in the semen of species with relatively small volumes of ejaculate, such as bulls, compared with those with a copious ejaculate such as the stallion. This may represent an adaptation to relatively anoxic conditions in the ejaculate of high sperm density, since anaerobic fructolysis can then take place. This appears to be of particular importance within the vagina, and species with vaginal insemination show high concentrations of fructose. Possibly in the less crowded conditions

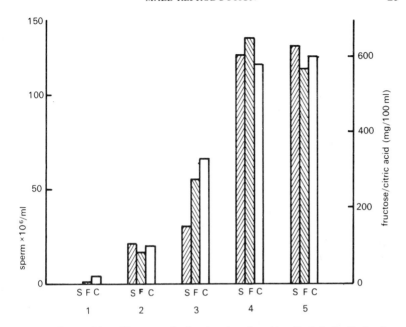

Figure 2.9 Composition of five successive fractions (numbered 1 to 5) of a bull split ejaculate. S, F, C = sperm, fructose and citric acid concentrations. (Data in Lutwak-Mann and Rowson (1953)).

within a large volume of ejaculate, sperm metabolism is more oxidative, and fructose correspondingly less useful.

A further peculiarity of the semen of species with a large volume of ejaculate is the high concentration of reducing substances such as the amino acid ergothioneine. In aerobic conditions, *in vitro* or *in vivo*, sperm produce potentially toxic amounts of hydrogen peroxide; ergothioneine

Table 2.1 Fructose and citric acid concentrations in human prostate and seminal vesicle secretions, and in semen. The figures are illustrative rather than precise, since concentrations vary widely; the figure quoted for citric acid in semen, for instance, represents the mean of observations which ranged from 96 to 1430 mg/100 ml. The samples of seminal vesicle fluid may have been contaminated with prostate fluid. The contributions of the prostate and seminal vesicle secretions to semen have been estimated at 13–33% and 46–80% respectively. All figures are in mg/100 ml.

	seminal vesicle secretion	prostate gland secretion	semen
citric acid	125	1500	376
fructose	315	0	224

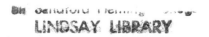

(and presumably other reducing substances) protects sperm against the action of hydrogen peroxide in species where aerobic conditions are experienced after copulation.

Numerous enzymes have been identified in semen; some of these are necessary for the clotting and subsequent liquefaction of semen which occurs after copulation in some species, or for the formation of vaginal plugs in others (see section 3.1). Other enzymes are concerned with a large number of other biochemical processes which take place in the semen after ejaculation, but the physiological importance of these is far from clear.

Prostaglandins occur in the semen of a number of species: human semen contains particularly large amounts, notably of prostaglandins E_1, E_2 and $F_{2\alpha}$. The prostaglandins are a family of fatty acids with a variety of important physiological functions (see figure 2.10). Prostaglandin $F_{2\alpha}$ is involved in several important aspects of reproduction (see also sections 3.4 and 7.2). One of its actions is to stimulate the contraction of smooth muscle; its presence in the semen may therefore accelerate sperm transport within the female tract (section 3.1). Other substances, probably gangliosides, with the ability to induce smooth muscle contractions are also present in the semen of some species. A protein with an apparently opposite action is secreted by the seminal vesicles of the bat *Myotis lucifuga*. This substance (which is as toxic as rattlesnake venom in most mammals) relaxes smooth muscle and may consequently be important in preventing the expulsion from the uterus of sperm stored there during hibernation. By manipulating leucocyte activity it may also protect the sperm from immunological attack by the female (section 9.2).

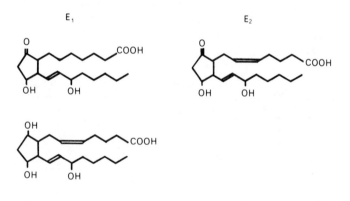

Figure 2.10 Structure of some prostaglandins of importance in reproduction.

Semen is therefore highly complex in composition and varies greatly from species to species. It is not possible to enumerate all of the constituents added by the accessory glands, still less to attribute a function to each. In fact, removal of one or other accessory gland does not necessarily impair fertility to a noticeable degree, and resuspension of washed sperm in a relatively simple physiological saline, followed by artificial insemination, can also result in conception. In some experiments, replacement of seminal plasma by saline positively enhances the ability of sperm to fertilize eggs *in vitro*. However, this does not mean that the composition of seminal fluid, and the idiosyncrasies shown by different species, are irrelevant, simply that their significance is quantitative rather than absolute. In some circumstances, the major function of semen—activation of the quiescent epididymal sperm and the provision of a suitable, if transient, suspension environment for them—can be performed by substances other than the normal ones.

2.7 Endocrine control of male reproduction

In outline, the endocrine control of the reproduction of male mammals conforms to the general scheme discussed in chapter 1. Steroid hormones produced by the testis control, besides spermatogenesis, the development and functioning of the accessory organs, secondary sexual characteristics, and sexual behaviour. In turn, the testis is controlled by the glycoprotein gonadotrophins of the anterior pituitary or adeno-hypophysis, in particular FSH or LH (Follicle Stimulating Hormone and Luteinizing Hormone; so named after their function in the female), while the anterior pituitary is under the control of the hypothalamic region of the brain.

The principal source of gonadal steroids is the Leydig tissue, also known as the interstitial cells (from which derives the alternative name of LH: ICSH, or Interstitial Cell Stimulating Hormone). These cells lie between the seminiferous tubules (see figure 2.3). Both androgens and oestrogen are produced, but the former generally predominate. The principal androgen secreted is testosterone (figure 4.6).

Dependence of testosterone secretion on pituitary LH is readily demonstrated. Hypophysectomy (removal of the pituitary) or the selective removal of LH from the circulation by combination with specific anti-LH antibody, is rapidly followed by a fall in circulating testosterone; injection of LH generally stimulates testosterone production. Prolactin appears to act synergistically with LH in stimulating steroidogenesis. In a hypophysectomized animal injected with prolactin and LH, blood testosterone levels are several times higher than when LH alone is injected.

Recent research has established the probable sequence of events in the Leydig cells. LH is taken up by specific receptor molecules at the cell surface; as a result, a membrane-bound adenyl cyclase enzyme stimulates the conversion of adenosine triphosphate (ATP) to 3′-5′-cyclic adenosine monophosphate (cyclic AMP, cAMP); this then acts on an inducer protein which promotes the formation of biologically active steroids from their precursor, cholesterol (figure 4.6). The action of cAMP as an intracellular "second messenger" to signal throughout the cell the presence of LH serves also to amplify the signal: one molecule of LH will cause the production of many cAMP molecules, each of which in turn leads to the formation of many steroid molecules.

LH may have some effects also on Sertoli cells. In general, however, its actions are restricted almost entirely to a single target cell type. Contrasting with this, the androgens have a multitude of effects on numerous target tissues, including various parts of the brain, secondary sexual characteristics (such as beard growth, and baldness in humans) and the sexual accessory organs.

Regression of the prostate and seminal vesicles follows castration; injection of testosterone reverses the process. Before the development of radioimmunoassay techniques (chapter 1) the sensitivity of these organs was exploited in bioassay determinations of androgen concentration. More recently, they have proved suitable for elucidating the mechanism of steroid action on target cells.

Unlike the gonadotrophins, steroids are small molecules and lipids; both properties enable them to diffuse freely in and out of cells through the predominantly lipid cell membranes. In cells which are responsive to steroids, specific receptor molecules are present in the cytoplasm (in concentrations of the order of 10^4 per cell), so that the steroid molecules pass in, combine with the receptor molecules, and remain within the cell. Radioactively labelled steroid is found to accumulate within the nuclei of target cells, which indicates that the next stage of the process involves interaction between the steroid-receptor complex and the genome, presumably involving the activation of specific genes.

Testosterone is not the only androgen to operate in this way to stimulate the growth, and maintain the secretion of the accessory glands. 5α-dihydrotestosterone (DHT) is highly active, particularly in maintaining the size of the prostate. Large amounts of a 5α reductase enzyme, which converts testosterone to DHT, are present within the prostate, as well as in nost other androgen target tissues. It appears that in most circumstances the biologically active androgen is actually DHT, and testosterone itself is merely a circulating precursor which must be converted locally to DHT before combining with receptor molecules. Yet another product of testosterone, 5α-androstanediol may also in some circumstances be the active agent. This suggests the possibility of modulation of the control of different male reproductive functions by alterations in the relative amounts of different androgens, or of the appropriate enzymes which produce them.

Androgen receptors also are present on Sertoli cells, and possibly also on spermatogonia and primary spermatocytes. Testosterone does seem to be essential for most stages of meiosis during spermatogenesis. It is not clear whether its effect is direct on the spermatocytes, or indirectly through the Sertoli cells. The hormonal requirements to restore spermatogenesis in the rat after oestrogen treatment have been worked out by Steinberger, and are summarized in figure 2.11.

FSH is also required, particularly during spermiogenesis. It may act directly on germ cells, but has now been clearly shown to stimulate Sertoli cells, inducing them to secrete into the tubule lumen a protein (Androgen Binding Protein, ABP) with a high affinity for testosterone and DHT. This may serve to detain some of the testicular androgen within the testis and prevent it from entering the whole body circulation. It can then distribute

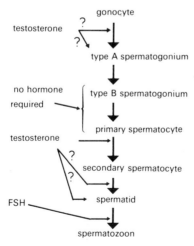

Figure 2.11 Probable hormonal influence on spermatogenesis. (From Steinberger and Duckett (1967)).

androgen to the germ cells from the luminal side of the seminiferous epithelium and also deliver it, in the seminal fluid, to the epididymis and accessory organs. Steroid binding proteins with similar affinities are present also in the circulation.

The hormonal treatment required to initiate spermatogenesis in an immature animal, or one whose testes have regressed after hypophysectomy, differs from that needed to maintain established sperm production. If a mature male is hypophysectomized and immediately treated with androgen, spermatogenesis will continue. No FSH is required. If 10 days elapse between hypophysectomy and the androgen treatment spermatogenesis does not recommence; FSH is now necessary. Testosterone stimulates ABP production (and consequently amplifies its own action) but can only do so with mature Sertoli cells previously acted upon by FSH. A current view of the endocrine control of the testis is summarized in figure 2.12.

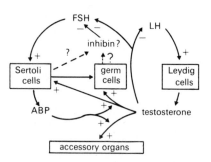

Figure 2.12   Hormonal control of male reproduction.

There has been considerable controversy over the question of whether FSH and LH are controlled by separate releasing factors or whether a single releasing factor is common to both. Some doubt remains, but it is now generally agreed that a single factor is probably responsible. This is known variously as LH releasing factor (LRF, LH–RF), LH releasing hormone (LH–RH), FSH/LH releasing factor (FSH/LH–RF) and so on. It is a decapeptide

(pyro) GLU–HIS–TRP–SER–TYR–GLY–LEU–ARG–PRO–GLY–NH$_2$

produced by neurosecretion in the ventro-medial region of the hypothalamus and passing to the anterior pituitary (adenohypophysis) through a portal blood system (figure 2.13). It appears to determine the rate

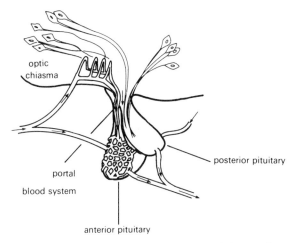

Figure 2.13 Relations between the pituitary and hypothalamus of a mammal.

of gonadotrophin synthesis as well as release, which may explain the observation that treatment of the pituitary with LRF increases its responsiveness to further LRF stimulation. Prolactin is similarly controlled, but by an inhibitory factor (PIF). A releasing factor controlling prolactin has also been postulated, but there is little good evidence for this.

LRF production and LH output are also controlled by a negative feedback mechanism from the testis. Castration of a ram is rapidly followed by a rise in serum LH; injection of testosterone depresses the LH concentration (figure 2.14). This indicates a fairly straightforward homeostatic mechanism, with LH stimulating the production of testosterone, which in turn acts back to suppress the production of further LH. The principal site of this testosterone action is the hypothalamus, and there is some evidence for androgen receptor sites there. As with the sexual accessory organs, testosterone may again not be the active androgen: the enzyme testosterone 5α-reductase is present in the hypothalamus, and the implantation there of small quantities of DHT suppresses LH production.

There is some evidence that testosterone may also alter the sensitivity of the pituitary to LRF. Conversion of testosterone to DHT can be carried out in the pituitary. The activity of the enzyme increases greatly after castration. Although the mechanism of this rise is unknown, its occurrence does introduce the interesting possibility that testosterone may in some way influence the rate of its own conversion to DHT.

Recently, it has become apparent that neither LRF, LH nor testosterone

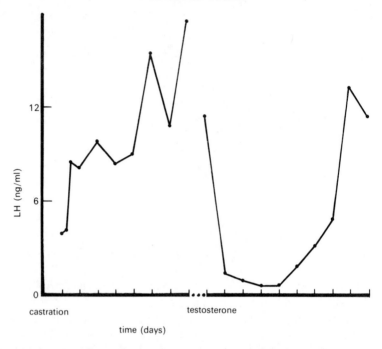

Figure 2.14 The effects of castration and subsequent androgen injection (400 mg testosterone propionate) on the level of circulating LH in the blood of a ram. Compare with figure 4.12. (From Short in Austin and Short (1972)).

are released continuously. Frequent sampling shows that all three have pulsatile patterns of release, the frequency of which may vary (figure 2.15). Rams changed from long to short day-length show a frequency of testosterone peaks of more than 10 per day, compared with less than 1 per day following the reverse alteration in daylight. Target organs, on the whole, probably act as integrators and respond to an average hormone level over a period of time. It is conceivable, however, that some information is transmitted within the body by variation in the frequency of hormone pulses, rather than their amplitude, or total hormone output.

In most circumstances, LH and FSH fluctuate together. This is not surprising, since they are controlled by a common RF. In some circumstances, however, the control of FSH may be divorced from that of LH. Injection of low doses of DHT into male rats suppresses LH only, although higher doses may suppress FSH as well. Conversely, the anti-androgen cyproterone acetate causes LH to rise, with little change in FSH

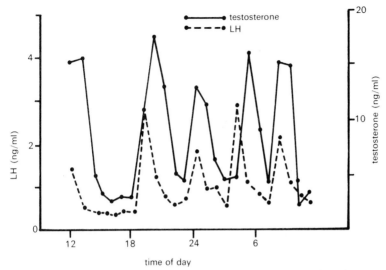

Figure 2.15 Variations in the level of LH and of testosterone in the blood of a bull. (From Katongole, Naftolin and Short (1971)).

levels. Evidently something other than androgen is involved in the control of circulating FSH levels.

One possible agent is oestrogen, also produced by the testis. Oestrogen can suppress FSH preferentially, but the results of oestrogen treatment are contradictory; in some cases LH is preferentially lowered, in others FSH and LH fall together.

Injection of cadmium salts causes sterility, with the seminiferous tubules virtually devoid of germ cells, but the endocrine capacity of the testis unimpaired. In these circumstances, LH levels are normal, but FSH significantly raised. This suggests that the hypothetical inhibitor of FSH—known as inhibin—is associated with germ cell and not Leydig tissue. Rete testis fluid and seminal plasma appear to contain a protein with the requisite properties, although it has not yet been completely identified.

On the other hand, rearing rats on a diet deficient in vitamin A results in tubules containing Sertoli cells with very few spermatogonia, and no other germ cells detectable. In these rats, FSH levels are within the normal range. Presumably inhibin levels are likewise unaffected, which suggests a Sertoli cell origin for inhibin. Similarly, the addition of a suspension of Sertoli cells (containing, however, a small proportion of residual germ cells) has been found to depress the output of FSH by cultured cells of the anterior pituitary. LH levels are unaffected. The substance which depressed FSH

c

secretion was identified as a heat-labile macromolecule, presumably a protein.

Therefore, although the evidence is in some respects conflicting, inhibin may be a protein, released by Sertoli or germ cells into the testis fluid, resorbed distally (possibly in the epididymis) and after passage in the circulation, acting directly on the pituitary to inhibit FSH release. If this turns out to be correct, it constitutes a means of monitoring and controlling rates of sperm production independently of androgen production.

A further complication of the control of FSH release is presented by the apparently variable properties of the hormone. At higher levels of testosterone, the FSH release is of greater biological activity, and of longer lifespan in the circulation, than at other times. (Ovarian hormones cause the opposite effect.) The significance of this control by steroids of the *quality* of a pituitary hormone, as well as its *quantity*, is not yet clear.

The posterior pituitary, in effect an extension of the brain rather than a separate organ, also produces hormones relevant to reproduction. Oxytocin (figure 1.1) causes smooth-muscle contractions; as it is released during coitus in humans (in both sexes) it may enhance the neural reflexes which expel semen during ejaculation. Vasopressin has also been implicated in sperm transport within the male genital tract.

Male sexual behaviour is to some extent under androgen control. Figure 2.16 shows the decline in sexual performance of male rats after castration and subsequent testosterone therapy. Unlike the action of testosterone on the brain in the control of gonadotrophin release, the behavioural effects appear to involve its conversion not to dihydrotestosterone (DHT) but to oestradiol, which then combines with intracellular oestrogen receptors. The extent to which sexual behaviour depends on testosterone, and diminishes after castration, varies according to species, and to the extent of sexual experience before castration.

If hormone concentrations affect brain activity and behaviour, it is also true that brain activity—specifically, sensory information—can affect hormone production. A bull shown a cow but prevented from mating ("teasing", as it is technically known among cattle breeders) responds with an unscheduled peak of LH, followed by a rise in testosterone. Evidence has been presented that anticipation of sexual activity in men can cause a rise in circulating testosterone levels, as measured by the rate of beard growth. Laboratory attempts to demonstrate correlation of sexual arousal with hormone fluctuations in humans have failed.

Other forms of environmental stimulus may also modify the male reproductive system. Some males are seasonal breeders, responsive to

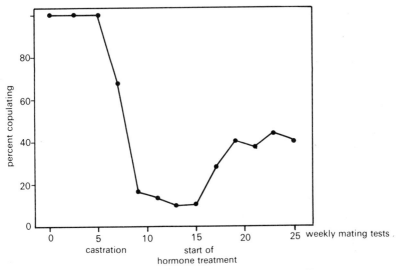

Figure 2.16 Effect of castration and subsequent androgen treatment (1 µg testosterone per day) on the sexual performance of male rats. (From Beach and Holz-Tucher (1948)).

daylength. Visual information, mediated in some way by the pineal body, can affect gonadotrophin production. The mechanism is not clear, but the pineal body is known to produce substances with anti-gonadotrophic properties, including the indole melatonin, and arginine vasotocin (figure 1.1). Pineal activity can in turn be modified by reproductive hormones, so this enigmatic organ may yet come to be regarded as an integral part of the reproductive system.

CHAPTER THREE

SPERM IN THE FEMALE TRACT

3.1 Sperm transport in the female genital tract

During copulation, semen is ejaculated into the genital tract of the female. The precise site of deposition varies from one species to another. In primates, rabbits, cows and cats, semen is deposited in the vagina; in ferrets, pigs and most rodents it is propelled directly into the uterus. The volume of the ejaculate and its sperm content also vary considerably. At one extreme, rats and mice ejaculate some 50,000,000—60,000,000 sperm in a compact mass around 0·1 ml in volume; at the other extreme is the pig, where the female may receive 250 ml of semen containing 5,000,000,000 spermatozoa.

Once ejaculation has taken place, the semen of many species coagulates to a greater or lesser extent. In rats and mice, a hard vaginal plug forms. This is shed within a few days, along with a large number of trapped sperm. A male rat whose seminal vesicles and coagulating glands (section 2.1) have been surgically removed will copulate readily, but no vaginal plug forms. Nor are any sperm found beyond the vagina. This indicates that the function of the plug may be to help in propelling semen into the uterus, and then to block the canal of the uterine cervix and so prevent the escape of the semen back into the vagina, and possibly to the exterior.

In humans and other primates, the semen coagulates within one minute of ejaculation and liquefies again after about 20 minutes. As the sperm becomes fully motile only after this, it may be a function of the coagulum to form a temporary reservoir of sperm, as well as reducing the chance of sperm loss from the vagina.

In some species, prolongation of copulation might serve the same purpose. In many carnivores, swelling of part of the penis causes it to become locked within the vagina; dogs may be unable to withdraw for 15 minutes or so. Grizzly bears may copulate continuously for up to one hour, and ferrets and mink for three hours; whereas a lioness copulates, on average, every 15 minutes for several days on end. Although there is a lack

36

of specific evidence, it seems likely that extended periods of copulation, whatever their other effects, will promote sperm retention.

Between ejaculation and fertilization, sperm must successfully negotiate the female genital tract. Those deposited directly in the uterus must survive the uterine environment, traverse the narrow uterotubal junction, pass up the isthmus of the oviduct (Fallopian tube), through the constriction between the isthmus and ampulla of the oviduct, and finally enter the ampulla, where fertilization may take place. In species with vaginal deposition of sperm, the vagina and cervix of the uterus represent additional obstacles (figure 3.1).

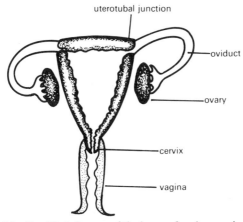

Figure 3.1 Simplified diagram of the human female reproductive tract.

The motility of human spermatozoa is totally inhibited at a pH lower than 6. Vaginal secretions in humans are acid, and may have a pH as low as 3–4, so that the vaginal environment is not a favourable one for sperm activity. Radiotelemetry has made it possible to monitor the pH of the human vagina continuously during coitus. At the moment of ejaculation, the vaginal pH rises sharply to 7·2 and remains there for some time (figure 3.2). This occurs only when semen is permitted to enter the vagina. One function of the seminal fluid, in humans at least, is apparently to reduce the vaginal acidity and so permit sperm motility.

The cervix of the uterus represents in many species a considerable barrier to passage of sperm into the uterus. Apart from its structure as a complicated series of narrow folds and crypts, it is occluded by the viscous

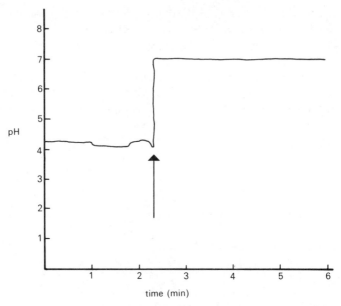

Figure 3.2 Change in pH within the human vagina during coitus, measured by radiotelemetry. The arrow represents the moment of ejaculation. (From Fox, Meldrum and Watson (1973)).

cervical mucus. The major fraction of this highly heterogeneous secretion consists of branched molecules of a glycoprotein; these are surrounded by a plasma which contains much protein (including globulins, albumin and enzymes) as well as lipids, simple sugars, and other substances. Bacteriocidal properties of human cervical mucus have been demonstrated, and it probably represents an essential chemical, as well as a physical, barrier to infection, since copulation is not a sterile process.

During most of the menstrual cycle, cervical mucus represents an equally effective barrier to sperm penetration. Scanning electron microscopy shows a dense mesh of fibrils, with gaps between them of only 0.5–2.2 μm (figure 3.3a). Shortly before the time of ovulation, and under the influence of the preovulatory peak of oestrogen (section 4.4), the pattern of mucus secretion changes. In humans, the amount produced daily rises from 20–60 mg to 700 mg; the water content of the mucus rises from 92–94% to 98%; the protein content decreases from 2.5% to 1.3%; and mucoids decline from 7% to 1.3%.

Along with these changes in chemical composition of the mucus are alterations in physical structure. The dense and impenetrable mesh gives way to a more open structure with much larger gaps (2–5 μm) between the mucus fibrils (figure 3.3b). The glycoprotein molecules become arranged in parallel micelles lying longitudinally with respect to the main axis of the cervical canal. On entering the mucus, spermatozoa tend to become orientated along these micelles, and actively-swimming sperm are in consequence guided upwards through the cervix (figure 3.4). Sperm passage through the cervical mucus is aided by proteoytic enzymes in the seminal plasma and by a decline, at mid-cycle, in the level of protease inhibitor present in the mucus.

The loose network of glycoprotein strands at mid-cycle has been shown by Nuclear Magnetic Resonance Spectroscopy (NMR) to be in a state of intense thermal oscillation. Mechanical resonance between the beat frequency of the molecular lattice of the mucus helps to propel sperm rapidly through the mucus. This will only apply to sperm whose tail beat frequency is normal, and thus coincides with one component of the mucus vibration. The cervical mucus appears to act as a selective filter which favours normally active sperm and tends to exclude all other objects. Temporarily permitting free access to spermatozoa at mid-cycle, when fertilization is possible, does not necessarily entail removing all barriers to infection.

A further consequence of the orientation of sperm along mucus strands is that not all sperm pass directly into the uterus. By following mucus strands to their point of origin, many sperm may accumulate within the cervical crypts. Here they are protected from phagocytosis, and possibly also from physical expulsion. Certainly dead sperm are confined to the central portion of the lumen of the cervical canal and are soon eliminated. The cervix may therefore be a sanctuary for the sperm and a reservoir from which, over a period of time, they may be released into the uterus.

In some respects the relationship of uterus to uterotubal junction parallels that of vagina to uterine cervix. Like the vagina, the uterus represents an environment potentially hostile to sperm. After copulation, there is generally a massive infiltration into the uterine lumen of leucocytes which phagocytically destroy any sperm which are still present there. Sperm which have attained the uterotubal junction are, in species such as the pig, protected from phagocytosis. The uterotubal junction, like the cervix, appears to act both as sanctuary and reservoir for the sperm; release of sperm into the oviducts over a period of time may be important in ensuring that sperm are continuously present at the site of fertilization.

The uterotubal junction acts also as a valve, regulating the entry of sperm

(a)

(b)

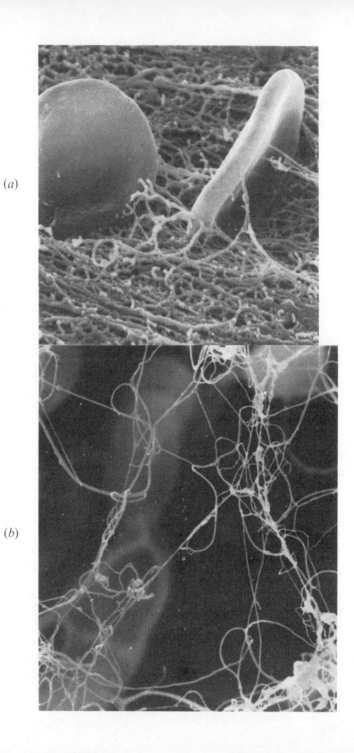

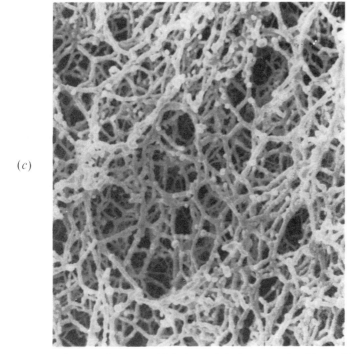

(c)

Figure 3.3 Baboon cervical mucus (a) during period of menstruation. Note red blood cells among the mucus strands (× 10,000) (b) on day of ovulation (× 5700) (c) towards end of menstrual cycle (× 10,400). (Photographs courtesy of Dr F. C. Chrétien).

into the oviducts. In the rabbit, it comprises a complex arrangement of folds, undergoing rhythmic contractions (figure 3.5). Physiological saline, injected under pressure into the lumen of the uterus, cannot pass into the oviduct; the uterus may even burst before the saline can be forced past the uterotubal junction. If the lower end of a rabbit uterine horn is ligated, the oviduct severed at the uterotubal junction, and sperm injected into the uterine lumen, periodic spurts of sperm release can be observed at the cut end of the junction. In the intact animal, sperm passage from the uterus is presumably regulated by similar periodic relaxation of the sphincter-like muscle of the junction.

In the rat (but not in the pig) dead sperm fail to pass the uterotubal junction. It may therefore, like the cervical mucus, serve as a selective filter with preferential passage of active sperm. The importance of this mechanism would presumably be greater in those species in which the

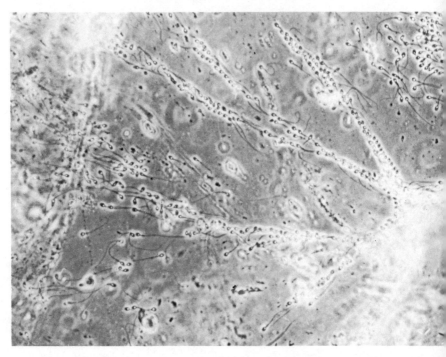

Figure 3.4 Human spermatozoa migrating through cervical mucus; note alignment along mucus fibrils. (From Parish, W. E. and Ward, A. (1968), *J. Obstet. Gynaecol. Brit. Cwth.*, **75**/11, 1089–1100 (fig. 6, p. 1096)).

cervix does not represent a significant barrier to sperm passage at copulation.

The female tract, then, contains a series of structures which in various ways restrict the passage of sperm. The effectiveness of these barriers is reflected in the distribution of sperm within the female tract after mating (Table 3.1) The sperm recovered from a particular part of the tract represent only the sperm currently in transit, and not the total number of sperm which reach that part of the tract. Considering the entire period for which sperm from one ejaculate will survive, it is probably more useful to think of the cervix and the uterotubal junction as reservoirs rather than barriers which few sperm can surmount; by detaining the bulk of the sperm, they ensure a continous supply of smaller numbers, over a period of time, to higher levels of the tract.

The time after ejaculation at which sperm first appear in the oviducts

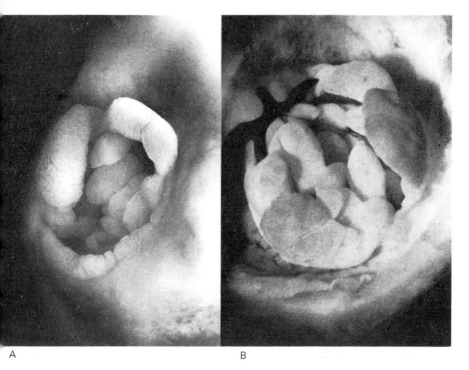

A   B

Figure 3.5 The appearance of the living uterotubal junction of rabbits. (A) is in a pubertal animal and (B) in an animal around the time of ovulation. (× 18) (From Hayashi, R. and Blandau, R. J. (1968), Gamete transport—comparative aspects, pp. 129–62 in Hafez, E. S. E. and Blandau, R. J. (eds.), *The Mammalian Oviduct*, (University of Chicago Press) figs. 10*a* and *b*, p. 144).

Table 3.1 Distribution of sperm within the female tract of a rabbit 10–11 hours after artificial insemination with sperm suspensions of two different concentrations. (The figures are antilogarithms of the logarithmic mean in each case.) (Morton and Glover (1974): *J. Reprod. Fertil.*, *38*, 139)

Sperm in inseminate (approx)	50×10^6	500×10^6
Numbers recovered from		
Vagina	759,000	4,470,000
Uterine cervix	398,000	1,200,000
Uterus	93,300	269,000
Oviducts	398	1,510

varies from species to species, but in most cases (the rabbit being a notable exception) transit through the female tract occurs within a few minutes (Table 3.2). The observed swimming rates of spermatozoa (less than 8 mm min^{-1}) are, in most species, simply insufficient to explain such rapid transport. The unimportance of the swimming movements of the sperm themselves is confirmed by the finding that even inert particles, deposited in the vagina of a sheep, are found in the oviducts within 15 minutes.

Although spermatozoan motility may affect rates of passage through cervical mucus and the uterotubal junction, the major propulsive forces are the contractions of the smooth muscle of the female tract. Hormone-dependent changes in intensity and frequency of contraction occur in the course of the oestrous or menstrual cycle in many species; it is reasonable to associate variations in contraction around oestrus with transport of sperm (and ova) which is likely to occur at this time.

Contractions of the female tract may be affected by many factors. In the sheep, stress delays the arrival of sperm in the oviducts. Conversely copulation in various ways aids sperm transport. In humans, radio-telemetry during coitus has indicated that the pressure within the uterus falls sharply at orgasm, becoming negative with respect to the vagina. Suction may therefore accelerate passage of sperm into the cervix. In cows, sheep and humans, physical stimulation of the cervix leads to the reflex release of oxytocin by the posterior pituitary, and this increases uterine tonus and amplitude of contractions. Finally, prostaglandins in the semen (section 2.6) may affect sperm transport in some species. This suggested role for seminal prostaglandin is not very firmly established since in some species, such as rabbits, prostaglandin levels in the semen are negligible, while in species like man with high seminal prostaglandin levels, only

Table 3.2 Time taken by spermatozoa of various species to pass through the female genital tract.

Mouse	15 minutes to upper oviduct
Rat	15–30 minutes
Guinea Pig	15 minutes
Rabbit	3–6 hours
Dog	2 minutes–several hours
Sheep	2–30 minutes to ampullae } separate estimates: trans- 4–8 hours to ampullae } port rate affected by stress
Cow	2–13 minutes
Pig	30 minutes in oviducts; 2 hours in upper half of oviduct
Human	30 minutes in oviducts; 68 minutes in ampullae

negligible amounts of seminal plasma pass with the sperm through the uterine cervix.

In the oviducts, gamete transport is more complicated, with the intriguing situation that sperm and ova apparently travel in opposite directions simultaneously. The ampulla and, to a lesser extent, the isthmus of the oviduct carry tracts of cilia which beat rapidly in the direction of the uterus. Fluid secreted by the cells of the oviduct moves in the opposite direction, since its passage into the uterus is prevented by the uterotubal junction, except for the brief period when the ova are permitted to enter the uterus. Some 90% of this fluid flows eventually into the peritoneal cavity. The interactions between these transport mechanisms and the peristaltic contractions of the oviduct have not been clearly elucidated, but it may be that sperm transport results mainly from the upward flow of oviduct fluid through the isthmus, while the broader lumen of the ampulla reduces the force of the upward current and facilitates downward propulsion of the ova by the cilia.

3.2 Physiology of spermatozoa in the female tract

After copulation, sperm enter an environment whose properties are determined by secretions of the female genital tract. Where the uterine cervix represents a significant barrier, this transition will take place as sperm enter the cervical mucus. The negligible amount of seminal plasma which enters the human uterus in normal coitus is demonstrated by the fact that artificial deposition of semen directly into the uterus causes specific discomfort. In species where semen is deposited in the uterus at coitus, the transition to a maternal environment will occur at the uterotubal junction.

Mammalian spermatozoa appear to contain enzymes necessary for carrying out the oxidation of fatty acids, glycolysis, the tricarboxylic acid (Krebs) cycle, electron transport system and oxidative phosphorylation, and possibly also the hexose-monophosphate shunt. Potentially, they are therefore capable of metabolizing a variety of substrates. While in a medium consisting largely of seminal plasma, the major substrates utilized are presumably the fructose and citric acid added to semen by the prostate and seminal vesicles, or their equivalents.

Human cervical mucus contains reducing substrates in amounts which vary throughout the cycle, but are comparable to those of seminal fluid. The most significant components are probably glucose, mannose and maltose. Of these, mannose and maltose decrease in concentration at mid-cycle, probably as a result of the greater water content of mucus at this time. By contrast, the glucose concentration rises at mid-cycle despite this dilution effect. Glycogen is also present, and is presumably made available to sperm following breakdown to glucose: an appropriate enzyme, found in cervical mucus, may have this function. The presence of spermatozoa in human cervical mucus *in vitro* doubles the rate of breakdown of sugar in the mucus; this indicates the probable importance of sugars to sperm *in vivo*.

Other substances in the mucus which might be utilized by the sperm are various amino acids (particularly tyrosine, phenylalanine and tryptophan), lipids and lipoproteins. However, since the motility of washed sperm is abolished in the absence of suitable sugars, and restored by the addition to the medium of sugars but not of amino acids, carbohydrate utilization appears to be of greater significance.

The composition of uterine fluid has been investigated in several species. In general, it appears as a watery secretion, variable in composition, but typically with significant amounts of reducing sugars present. The oxygen tension, estimated to range from 25 to 45 mmHg, is sufficient to support aerobic metabolism of the sperm. One particular feature of interest is the secretion by uterine cells of a diesterase capable of hydrolysing glycerylphosphorylcholine (GPC). This may explain the presence of GPC in relatively enormous amounts in the semen of species such as the rat (see section 2.5) as a nutrient available to sperm only after copulation and entry of semen into the uterus. However, GPC may have other functions, since the diesterase is found in some species in which seminal plasma does not accompany spermatozoa into the uterus, and is not found in all species with significant levels of seminal GPC.

The volume of oviducal fluid secreted varies in relation to the oestrous or menstrual cycle, with a rise in secretion rate around the time of oestrus (figure 3.6). Large amounts of carbohydrate are present in the fluids of several species studied, principally glucose (approximately 300 μg ml^{-1} in sheep oviduct fluid level at oestrus), although fructose, pyruvate, citrate and lactate are also present. Amino acids and fatty acids are also found in varying amounts. Since oxygen tension in the oviduct is relatively high (45–60 mmHg; 1–2 mmHg is sufficient to support sperm activity), the metabolism of sperm in the oviduct probably resembles that elsewhere in the female tract.

Several factors in the oviducal secretion have stimulating effects on sperm metabolism *in vitro*. Oviduct fluid of the ewe stimulates respiration of sperm in the absence of glucose. When glucose is present, this does not happen, but both aerobic and anaerobic glycolysis rates increase. The

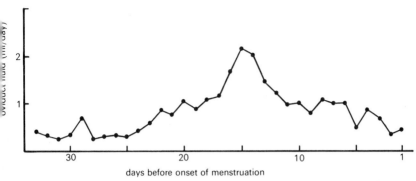

Figure 3.6 Secretion of fluid by the oviduct of a rhesus monkey in the course of the menstrual cycle. (From Yoshinaga, Mahoney and Pincus (1971)).

stimulatory factor has been identified as bicarbonate, although a non-dialyzable heat-labile factor has also been implicated in the rabbit. Washed human sperm respire at a rate approximately $1·4$ μl $O_2/h/10^8$ sperm. If bicarbonate is added, at a concentration of $3·54 \times 10^{-3}$g l^{-1} (lower than the concentration in oviducal fluid) the rate of oxygen consumption increases tenfold.

The composition of oviduct fluid, as well as that of uterine, cervical, and vaginal secretions, is highly complex, and varies considerably between species, between individuals of a species, and in different reproductive states. The significance to the sperm (if any) of most of the constituents has yet to be resolved.

Whatever the relationship between the sperm and the secretions of the female tract, it is generally a brief one. Table 3.3 gives the estimated

Table 3.3 Retention of motility and fertilizing capacity of spermatozoa in the female tract of mammals

	Maximum retention of	
	fertilizing capacity *(hours)*	*motility* *(hours)*
Mouse	6	13
Rat	10–14	17
Guinea Pig	21–22	41
Sheep	30–48	48
Cow	28–50	96
Horse	144	144
Human	28–48	48–60
Bat (various species)	135 days	149 days

maximum time for which sperm remain motile, and for which they retain the ability to fertilize ova. Estimates of the latter are probably less accurate, since sperm fertility is more difficult to assess.

A striking feature of the table is the exceptionally long survival of sperm in certain species of bat. In these animals copulation occurs in the autumn, and the sperm are retained in the uterus or oviduct until spring, when ovulation and fertilization take place. Why this should be necessary is not clear. The physiological mechanisms of long-term sperm storage have recently been investigated, but are still far from being fully elucidated. Glycogen and fructose are known to be secreted by the uterus during sperm storage, and may supply sufficient nutrient for prolonged survival, particularly in the low body temperature during hibernation. Apart from nutrition, one of the many points of interest in this unique mammalian situation is the question of how sperm survive in an immunologically hostile environment. This point is discussed in relation to more conventional mammals in section 9.3.

3.3 Capacitation and fertilization

Before making contact with ova to achieve fertilization, the spermatozoa must first pass through the several layers which invest the eggs (figure 3.7). First is the cumulus oöphorus, a mass of loosely-packed cells derived from the granulosa cells of the follicle which surrounded the egg before ovulation. (*Cumulus oöphorus* means "egg-bearing little cloud", an apt description.)

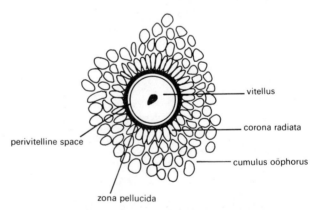

Figure 3.7   A mammalian ovum shortly after ovulation, showing the surrounding cells of the cumulus oophorus and corona radiata.

Within the cumulus lies the corona radiata which consists of follicle cells attached directly to the third layer, the zona pellucida. After passing through the zona, the sperm come into the perivitelline space, attach to the vitelline membrane, and so enter the egg. Acrosomal enzymes facilitate passage of sperm through these investing layers and, associated with the release of these enzymes, the sperm undergoes an "acrosome reaction" characterized by the apparent partial detachment of the acrosome from the sperm head.

Freshly ejaculated sperm are incapable of undergoing the acrosome reaction, and of penetrating eggs and achieving fertilization. This is true of attempts at fertilization both *in vivo* and *in vitro*. A period within the female genital tract appears to be necessary before full fertilization capacity will develop. The process undergone by the sperm within the female tract is known as "capacitation": capacitated sperm show greatly increased motility, beating their flagella in the characteristic "whiplash" pattern (figure 3.8), and can undergo the acrosome reaction, penetrate the

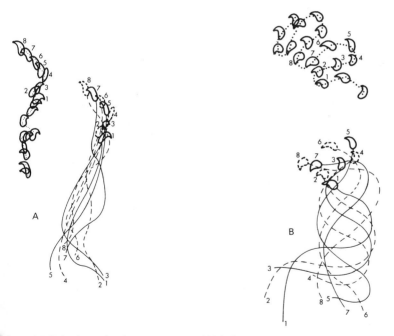

Figure 3.8 Behaviour of rodent spermatozoa (A) before and (B) after capacitation. These pictures represent tracings of the positions of individual spermatozoa, taken from a film at 0·02–second intervals. (From Yanagimachi (1970)).

D

surrounding layers, and fertilize ova. Uncapacitated sperm show none of these properties. The significance of capacitation for penetration of the egg is shown by the fact that even uncapacitated sperm can penetrate an egg from which cumulus, corona and zona have been artificially removed.

Different times are required for capacitation in different species: about 5 hours in the rabbit, probably 7 hours in the human, between $3\frac{1}{2}$ and $11\frac{1}{2}$ hours in the ferret, and 2 to 3 hours in the rat. Capacitation in the mouse is very rapid: too brief to determine precisely.

Capacitation is a reversible process. Seminal plasma contains at least one substance with a decapacitating effect on already capacitated sperm, suppressing the ability to undergo an acrosome reaction. Decapacitating Factor (DF) appears to be a glycoprotein, and it has been proposed that it is normally adsorbed on to the sperm during passage through the male tract before ejaculation, and lost or removed in the female tract after the sperm leave the environment of the seminal plasma. The uterus appears to have considerable ability to remove or destroy DF from sperm. In one experiment, even adding DF equivalent to 40 ejaculations to ejaculated sperm in the uterus of a rabbit failed to prevent capacitation.

The mode of action of DF is not clear, but at least one aspect of capacitation is a process of labilizing the acrosome so that the penetrating enzymes may be readily released on contact with the cumulus; it follows that decapacitation would involve stabilization of the outer acrosome membrane. There is good evidence for the uptake by the sperm, during passage through the male tract, of substances from the seminal fluid, and for the loss of some surface components after entry into the female tract. Direct evidence for a membrane-stabilizing role for any such substance is, however, still lacking; for example, no morphological changes in the acrosomal membrane are apparent accompanying capacitation or decapacitation. The processes leading up to fertilization are complex, and other steps besides membrane labilization might be blocked by Decapacitation Factors. One suggestion is that capacitation involves the activation of the acrosomal enzymes. Decapacitation could then represent an inhibition of enzyme activity. A seminal decapacitation factor has been found which does in fact inhibit one of the acrosomal enzymes, Corona Penetrating Enzyme (see below).

Many investigations have been carried out *in vitro* to determine the conditions necessary for capacitation. One feature which emerges clearly is that the requirements for *in vitro* capacitation differ according to species. Rabbit sperm, to become capacitated, require a highly complex environment, and the presence of live cells. With hamster sperm, contact between sperm and maternal cells is not necessary, but the culture medium must still be a highly complex one. Guinea pig sperm may become capacitated even in a saline medium lacking any macromolecules. This may indicate that capacitation following normal insemination is largely an endogenous process which, in an appropriate medium, will proceed spontaneously, although external conditions can certainly facilitate or inhibit it. How relevant *in vitro* capacitation is to the *in vivo* process is not clear.

Such capacitation has no apparent morphological correlates; the only evidence of its completion is the occurrence of an acrosome reaction and the concomitant increase in motility. This has made it difficult to separate the process of capacitation from its consequence, the activation of the sperm. Recently, however, Yanagimachi has shown that sperm which have been incubated in a calcium-free medium fail to show either increased motility or an acrosome reaction. If, after several hours, calcium ions are restored to the medium, the changes in appearance and behaviour associated with sperm activation can be elicited without delay, showing that capacitation has already occurred. It is therefore possible to separate capacitation from sperm activation and to show that the former is a largely autonomous process and the latter requires specific induction. Yanagimachi interpreted his experiments as demonstrating that two substances, present in bovine follicular fluid, could evoke the Ca^{2+} -dependent changes in hamster sperm. One, heat-stable and dialysable, promoted the "whiplash" motility which normally marks the completion of capacitation, while the other, non-dialysable and heat-labile, induced the acrosome reaction. This interpretation is not, however, universally accepted. Capacitation remains a somewhat mysterious and ill-defined process.

The sequence of events of the acrosome reaction is known in some detail (figure 3.9). As penetration of the cumulus begins the outer acrosome membrane and the plasma membrane of the spermatozoon fuse , and many small apertures appear in the resulting composite membrane. Through these, enzymes are released. As the sperm passes through the corona and reaches the outer surface of the zona, the fused plasma and outer acrosomal membrane, now highly vesicular in appearance are discarded. The sperm attaches to the zona: possibly specific receptor sites are involved, since sperm binding to the zona can be blocked either by coating of the zona surface with wheat germ agglutinin or treating it with pancreatic trypsin. The sperm now makes a narrow slit in the zona and passes through, typically following an oblique path, into the perivitelline space. Attachment of the residual plasma membrane to the surface of the vitellus may also involve specific receptor

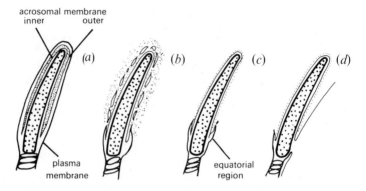

Figure 3.9 Events of the acrosome reaction. (*a*) intact sperm head (*b*) fusion of plasma membrane with outer acrosomal membrane, release of hyaluronidase, etc. (*c*) penetration of zona pellucida; fused plasma and outer acrosomal membrane discarded, exposing acrosin attached to the exposed inner acrosomal membrane. (*d*) attachment to egg, fusion of plasma membranes of egg and sperm.

sites. Fusion of the plasma membranes of sperm and egg is followed by dissolution of the cortical granules at the egg surface, entry of the sperm pronucleus into the egg cytoplasm, and eventually pronuclear fusion.

Sequential release of acrosomal enzymes occurs during penetration by the sperm. Many enzymes have been identified, although the function of some of these remains obscure. The first whose effect is detectable is a hyaluronidase. About half of the sperm hyaluronidase in intact rabbit sperm can combine with antihyaluronidase antibody; this indicates that much of the enzyme is exposed on the surface of the sperm. The remainder is presumably located within the acrosome, whence it is readily released both by the normal acrosome reaction and by artificial means. Since hyaluronic acid is a major component of the intercellular matrix which holds the cumulus cells together, sperm hyaluronidase is responsible for the passage of the sperm between the cumulus cells The scything action of the heads of rodent sperm after activation (figure 3.8) may help passage between the loosened cells. Passage through the corona radiata is similarly achieved by a Corona Penetrating Enzyme, which appears to be inhibited by a Decapacitating Factor (see above).

Penetration of the zona is achieved largely by the protease known as acrosin (or acrozonase). This probably closely resembles pancreatic trypsin, as zona lysis can be achieved artificially by trypsin, and specific inhibitors of pancreatic trypsin also block zona penetration. The very narrow track left by the sperm in the zona suggests that acrosin is probably tightly attached to the inner acrosomal membrane.

Other protease enzymes may be involved in addition. An acid protease (pH optimum around 3.5) has been identified in acrosomal material, and it has been suggested that the localized action of the sperm in the zona matrix creates an acid microenvironment in which acrosin itself with a pH optimum around 8, is no longer effective.

Neuraminidase is also present in the acrosome. Neuraminic acid is a component of the zona material, and it has been shown that *in vitro* treatment of rabbit zonae with sperm neuraminidase makes them more resistant to digestion with trypsin. Greater resistance to trypsin digestion is a feature of the zonae of fertilized ova. Similarly, *in vitro* treatment with bacterial neuraminidase inhibits sperm penetration. The obvious conclusion is that sperm neuraminidase has some role in the so-called block to polyspermy— the prevention of multiple sperm entry into an egg by the first sperm to arrive. Unfortunately, in the rabbit some 20–60 sperm contrive to pass through the zona, so the block to polyspermy in this species must lie at the surface of the vitellus.

In other species, a change in the penetrability of the zona does appear to be an important component of the block to polyspermy. When fusion of sperm and egg begins, one of the earliest morphological changes in the egg surface is the dissolution of the cortical granules and the release of their contents into the perivitelline space. The most direct evidence that cortical granule breakdown is implicated in the reduced penetrability of the zona comes from the observation that in aged eggs, failure of the cortical granules to discharge their contents is accompanied by the absence of "zona reaction".

Three explanations have been proposed of how the contents of the cortical granules might act. Firstly, they might have a "tanning" effect on the zona, altering its structure so that the zona material could no longer be digested by acrosin. Secondly, a specific inhibitor of acrosin might be involved. Finally, there is evidence that, in the hamster, one product of cortical granule breakdown is itself a trypsin-like protease very similar to acrosin. This might destroy specific receptor sites for sperm on the zona as well as at the surface of the vitellus, and so could explain the development of blocks to polyspermy at both levels.

CHAPTER FOUR

REPRODUCTION IN THE FEMALE: THE OESTROUS CYCLE

4.1 Reproduction in the female

It is generally accepted that the female mammal differs from the male in being cyclic in its reproductive processes. This is not entirely true. In the wild, the females of many species become sexually receptive once a year; if they fail to conceive, sexual receptivity does not recur until the subsequent breeding season. Such species (e.g. the fox *Vulpes vulpes*) are referred to as *monoestrous*. In other species, a failure to conceive is soon followed by a return to a receptive state. In some cases, many such cycles may recur within a single breeding season, or they may be continuous throughout the year. Species with recurring cycles are known as *polyoestrous*; a good example of a polyoestrous species is the mouse. However, such repeated failure to conceive is highly unusual in the wild. An uninterrupted sequence of oestrous or menstrual cycles is an artefact of the conditions in which laboratory animals are maintained, or of recently developed human attitudes to human reproduction.

The female reproductive cycle is in any case a combination of several

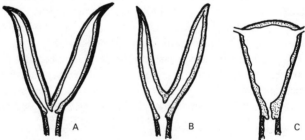

Figure 4.1 Some forms of mammalian uterus. (A) *duplex*: uteri separate, with separate openings into the vagina (most rodents, rabbit). (B) *bicornuate*: two uterine openings, a single cervical canal (guinea pig). (C) *simplex*: a single uterus formed by the complete fusion of the primitive paired uterine horns (primates). The *bipartite* uterus of carnivores is intermediate between (B) and (C).

different cycles. The central feature is the periodic maturation and ovulation of eggs which can be fertilized. In parallel with this ovarian cycle is the cyclical behaviour of the uterus and associated structures, so that, if fertilization does occur, the uterus is in a receptive state for implantation and gestation. Finally, the female of most species will accept a male only for a brief period around the time of ovulation (behavioural oestrus or "heat") when conception is most probable. At other times, she is uninterested or actively hostile. As in the male, the three separate classes of reproductive function, concerning the gonads, the genital tract and the brain, are controlled and coordinated largely by hormonal means.

4.2 The female genital tract

The morphology of the female reproductive tract varies. In most species ova are released from the ovary into the peritoneal cavity, picked up by a fimbriated funnel, and thus conveyed into the proximal end of the oviduct, while in others (mice and rats among them) the funnel forms an almost complete capsule enclosing the ovary. The distal end of the oviduct joins the uterus at the uterotubal junction. In many animals (particularly the guinea pig) this is an elaborate structure, muscular and with many complex folds of tissue, and acts as a valve to regulate entry of sperm into the oviducts. The human uterotubal junction, in contrast, is simple, lacks a sphincter, and does not significantly impede passage from the uterus. The relation of oviduct to gamete transport is discussed in chapter 3.

The uterus may be simple, double or intermediate in form. Simple uteri are found in primates, double in most rodents, and intermediate forms in ungulates and carnivores (figure 4.1). Distally the uterus connects with a vagina through a narrow, frequently highly complex region known as the *uterine cervix*. In primates the cervix is particularly well developed. Glandular tissue in the cervix produces a mucus secretion (chapter 3) which occludes the lumen and prevents entry of dangerous microorganisms. Around the time of ovulation the consistency of the mucus alters, and sperm can pass into the uterus and crypts of the cervical mucosa. Species in which insemination is uterine (e.g. mouse, pig) have a cervix which does not represent a significant barrier to sperm passage, and generally a correspondingly well-developed uterotubal junction.

The vagina also varies in structure according to species. In sheep and cattle, where copulation lasts for only a few seconds, the vagina is relatively simple. At other extremes, in many carnivores, the vagina contracts and locks on to the penis; in these species copulation may last for a considerable period. The vagina of the sow forms spiral grooves which match the ridged "thread" of the boar's penis and ensure a close fit during copulation, presumably to prevent a backflow of the copious ejaculate.

The structure and activity of the female tract changes cyclically. During the "follicular" phase of the oestrous cycle, when a crop of oocytes is undergoing maturation before being ovulated, the lumen of the uterus is narrow, and its epithelial lining low and cuboidal in form. The glands which lie in the endometrial mucosa are only slightly branched, and are inactive, and the mucosa itself contains many leucocytes. As ovulation and behavioural oestrus approach, cell division in the endometrium increases and the leucocytes vanish, while the lumen (in some species at least) greatly distends as a result of fluid secretion by the uterus and oviducts (figure 3.6). After ovulation, during the "luteal" period (when the evacuated follicles of the ovary have converted into corpora lutea: see next section), the endometrium thickens, its epithelium becomes columnar rather than cuboidal, and the mucosal glands extensively branched and coiled, and actively secretory. Physiological changes accompany morphological ones (figure 4.2).

Towards the end of the luteal phase, the endometrium again becomes infiltrated by large numbers of leucocytes, and reverts to its original state. Primates slough off much of the endometrial layer, together with some blood lost from the uterine blood vessels. This is known as *menstruation*. Other species also slough off the uterine lining, but since only the epithelial layer is involved and no blood is lost, the process is much less conspicuous. In cows and pigs, the endometrial lining is shed late in the luteal phase of the cycle, and in sheep early in the following follicular phase. Rats show almost continuous loss of endometrial cells. In mice and rats the corpora lutea do not become

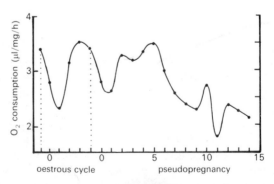

Figure 4.2   Oxygen consumption of slices of uterine tissue during the oestrous cycle and pseudopregnancy in the rat. (From Yochim 1975)

fully functional in the course of the infertile oestrous cycle, and persist for only a short period. There is therefore no clear separation of the cycle into distinct follicular and luteal phases (section 4.3).

Similar changes take place in the vagina. Since this is more accessible than the uterus, the cell contents of vaginal smears, reflecting changes in vaginal cytology, provide a useful means of diagnosing the oestrous state of an animal such as a mouse or rat. The conventional stages into which the rat cycle is divided are as follows:

early oestrus (ca. 18 h): vaginal smear mainly composed of (basophilic) nucleate epithelial cells
oestrus (ca. 25 h): epithelial cells (acidophilic) cornified and enucleate
metoestrus (ca. 5 h): some basophilic cells present
dioestrus (48 + hours): large numbers of leucocytes present
late dioestrus (ca. 7 h): leucocytes and some nucleate epithelial cells again present.

In a polyoestrous animal such as the rat, this cycle will probably be immediately followed by another one; the female of a monoestrous cycle enters a prolonged period of anoestrus, with the uterus and other parts of the genital tract in a regressed state.

Removal of the ovaries results similarly in regression of the uterus and a state of permanent anoestrus. It is possible, by treating an ovariectomized animal with steroids, to find out what combination is necessary to substitute for the missing ovaries. Specific oestrogen-binding receptors are present in the cytoplasm of uterine cells (see chapter 2 for discussion of the action of steroid hormone receptors). After combining with these, oestrogen induces an increase in uterine blood flow, and an increase in cell division rates in both endometrium and myometrium (the layer of smooth muscle surrounding the endometrium). In mice and rats treated in this way, water accumulates in the lumen and distends it, but in other species the uterine tissue becomes oedematous.

Oestrogen treatment also causes a rise in the concentration of cytoplasmic progesterone receptors, thereby increasing the sensitivity of the tissue to progesterone. Small amounts of progesterone applied to an oestrogen-treated uterus cause the endometrium to thicken, mainly as a result of an increase in coiling and convolution of the uterine glands. This sequence of hormone treatment—oestrogen, followed by progesterone and oestrogen together—crudely mimics the steroid exposure of the uterus during an oestrous cycle in the intact animal. Oestrogen secretion rises to a peak shortly before ovulation (e.g. figure 4.7) while progesterone is secreted in small amounts before ovulation, increasing as the corpus luteum becomes fully functional.

Progesterone appears to have a further effect on the uterus: it depresses the endometrial content of progesterone receptors. This may explain why in many species the luteal phase terminates, and the uterus regresses, with no apparent external influences.

4.3 The ovary

Gamete production, in the female as in the male, requires both cell multiplication and meiosis. Figure 4.3 illustrates the main stages of oogenesis for comparison with spermatogenesis (figure 2.11). The most striking difference is that in the female mammal germ cell multiplication has ceased by the time of birth, and only meiosis remains. All germ cells which may eventually be ovulated are, in most mammals, primary oocytes prior to the first division of meiosis. After puberty, a proportion of these enter a growth phase at each cycle and may finally be ovulated. The completion of meiosis in general occurs only after sperm penetration.

The ovary of a human female fetus in the sixth month of pregnancy contains more than 6×10^6 germ cells. By the time of birth, the number has fallen to around 2×10^6, and by puberty only some 3×10^5 remain. Of these,

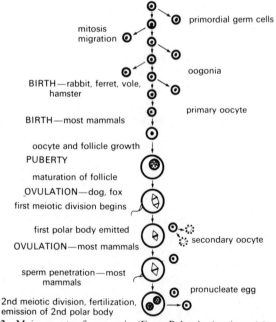

Figure 4.3   Major events of oogenesis. (From Baker in Austin and Short 1972)

no more than about 400 could possibly be ovulated; nevertheless, by the time that menstrual cycles cease, the oocyte supply of the ovaries is totally depleted. Similar germ cell losses have been shown in many species; even in such eccentric animals as the plains vizcacha (*Lagostomus maximus*) which ovulates as many as 800 eggs at each cycle. Oogenesis, no less than spermatogenesis, involves massive losses of germ cells (see section 2.3).

The kinetics of oogenesis has been intensively studied in a number of species, particularly the mouse. A mature mouse ovary contains a large pool of undeveloped oocytes in a quiescent state. These are less than 20 μm in diameter; any oocyte larger than this has started development. From the time of sexual maturity, oocytes are continually leaving this pool and increasing in size. The frequency with which oocytes start to develop does not vary greatly from day to day, and is affected neither by blocking the secretion of endogenous gonadotrophin nor by the injection of exogenous gonadotrophin. The rate at which oocytes leave the resting pool does, however, appear to be determined by the total size of the pool (figure 4.4)

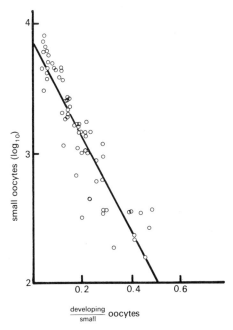

Figure 4.4   Relationship between the number of oocytes which have started development and the number remaining within the pool of those which have not started to develop, in the ovary of the mouse. (From Krarup, T., Pederson, T. and Faber, M. (1969)), Regulation of oocyte growth in the mouse ovary, *Nature* **224**, 187–8. (fig. 1, p. 187).

although it is not clear how this control is exercised. Degenerating follicles may also release a substance which lowers the initiation rate. Whatever factors modulate the frequency of initiation, the process itself is clearly a continuous one, in marked contrast to the discontinuous and periodic manner in which mature oocytes are ovulated.

Once started, development proceeds inexorably and can end only with ovulation or degeneration (atresia). In many species, the time required to develop to the preovulatory stage is 10–17 days. The oocyte itself does not increase greatly in size; the most striking feature of development is the organization and rapid growth of the tissues of the surrounding follicle (figure 4.5). This consists of an inner layer known as the *membrana granulosa* and the outer layers of the *theca interna* and *theca externa*. During the last few days of follicle development, the rate of proliferation of granulosa cells increases greatly; spaces appear between granulosa cells and merge into a large fluid-filled antrum, and the volume of the follicle increases. This culminates in the mature, or Graafian, follicle. Once the oocyte has been expelled at ovulation, the follicle collapses; blood vessels penetrate the basement membrane and vascularize the granulosa cells for the first time; and further proliferation of the granulosa cells produces a structure known as the *corpus luteum* (plural: corpora lutea).

In the course of the oestrous cycle the ovary produces steroid hormones, chiefly oestrogens and progestogens (figure 4.6), together with some androgen. Cells of the theca interna are the main source of oestrogen and androgen, and granulosa cells of progestogen; *in vitro* studies indicate that the difference between the performances of two cell types is quantitative rather than qualitative. Until ovulation and luteinization, the granulosa cells lack vascularization, and can obtain nutrients and oxygen only by diffusion through the basal membrane which separates them from the theca. This presumably limits their capacity for steroid synthesis. The restraint is lifted during luteinization, and the vascularization of the granulosa accounts for the increase in progestin production which occurs then. In some species, the rat being an example, the interstitial tissue of the ovary can also produce large amounts of progestogen. The preovulatory peak of progestogen in the rat is due to interstitial production.

Both gametogenic and endocrine functions of the ovary are under pituitary control. Follicular development, ovulation and luteinization do not occur after hypophysectomy. Developing follicles probably require only a low, and continuous, stimulation with FSH to continue their development, or at least to prevent atresia. If mice are given an injection of Pregnant Mare Serum Gonadotrophin (PMSG; similar in its effects to FSH) the proportion of large

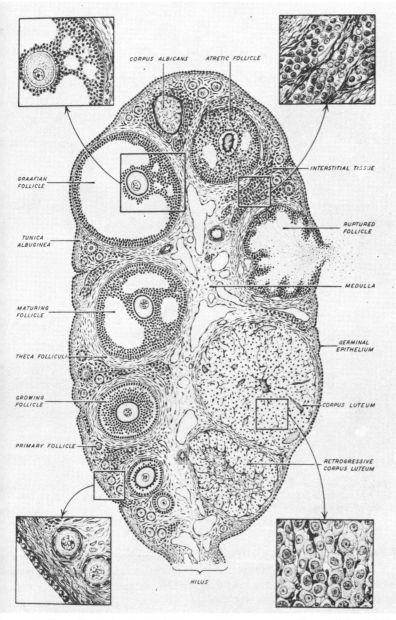

igure 4.5   Generalized representation of a mammalian ovary. (From Turner, C. D. (1966)), *General Endocrinology*, 5th ed. (W. B. Saunders Co.) (fig. 12–7, p. 400).

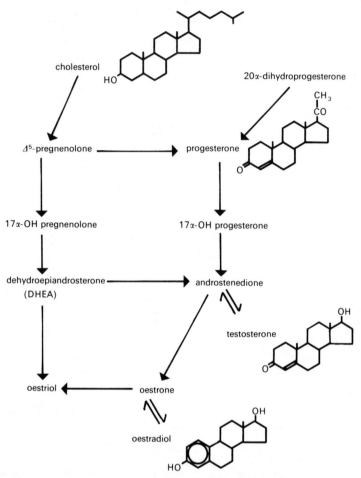

Figure 4.6 The major pathways of gonadal steroid synthesis. The route from pregnenolone to androstenedione via progesterone is known as the Δ^4 pathway, and that via dehydroepiandrosterone (DHEA) as the Δ^5 pathway. These pathways occur preferentially within the granulosa/corpus luteum and theca respectively.

follicles in the ovary found to be atretic falls to approximately 33% from the normal figure of 76%. The total number of large follicles is not altered by this treatment, so PMSG allows the continued development of follicles which would otherwise have become atretic; it does not increase the amount of developing follicles. A similar effect explains the multiple births which often follow treatment of women with "fertility drugs" (see section 10.6).

After the completion of follicle maturation, a sharp rise in output of LH by the pituitary is followed by ovulation. A possible explanation of how LH acts to cause a follicle to ovulate has recently been formulated. This is based on the finding that, contrary to earlier suggestions, the Graafian follicle does not simply accumulate fluid until hydrostatic pressure causes it to burst. In fact, the intrafollicular pressure actually declines slightly before rupture of the follicle wall. Rupture probably therefore results from an increase in distensibility and a reduction in breaking strength of the follicle wall. The increase in distensibility can be duplicated *in vitro* by treatment with LH, cAMP, progesterone or collagenase. LH appears to combine with specific receptors on the surface of the follicle cells, stimulating adenyl cyclase, cAMP production, and progesterone output. Progesterone then induces synthesis or activation of a collagenase enzyme which, acting on the collagen framework of the follicle wall, increases distensibility and lowers breaking strength; rupture of the wall and release of the oocyte follow.

The oocyte appears to have an inhibitory action on luteinization of the follicle, since extirpation of the oocyte from a preovulatory follicle is followed by premature luteinization and progesterone production. Normally, this inhibition would be removed by ovulation. LH also exerts a direct luteinizing effect, since LH treatment can overcome the inhibition and luteinize an unovulated follicle. The mode of action of LH is not clear, but it does increase the content of several steroidogenic enzymes in the whole rat ovary.

Recent work on the cellular action of the gonadotrophins on ovarian follicles has demonstrated the subtlety of the interactions between hormones in the control of oocyte maturation. The sensitivity of a target tissue is determined in part by its content of receptors for the hormone which affects it. FSH is found to stimulate the production of FSH receptors on granulosa cells; prolonged exposure to FSH will therefore progressively increase sensitivity to FSH. Similarly, intracellular oestrogen receptors increase in concentration with oestrogen treatment. Oestrogen and FSH together appear to stimulate production of LH receptors, and therefore render the follicle sufficiently sensitive for LH to achieve ovulation and luteinization.

For a freshly formed corpus luteum to achieve its full function and to continue to produce progesterone, further hormonal maintenance is generally required. In the rat, the corpora lutea do not become fully functional during an infertile oestrous cycle, and only low levels of progesterone are secreted (figure 4.7). In circumstances where the corpora lutea do become functional (see chapter 5) this is due to the action of pituitary hormones with a luteotrophic effect. The principal luteotrophic hormone in the rat is prolactin; suppression of prolactin release prevents any rise in

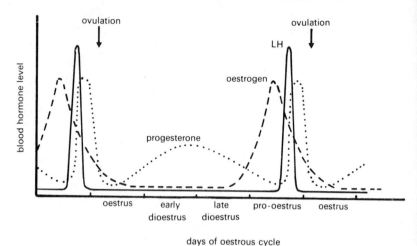

Figure 4.7   Hormonal fluctuations in the course of the rat oestrous cycle. (From Short in Austin and Short 1972)

progesterone output by rat corpora lutea. Injection of prolactin, conversely, can stimulate the formation of fully active corpora lutea. Prolactin is known to inhibit a luteal enzyme, 20 α-hydroxysteroid dehydrogenase, which converts progesterone to the relatively inactive metabolite 20 α-dihydro-progesterone (figure 4.6). LH and oestrogen may also be luteotrophic in some circumstances, and it is possible that the interaction between these three hormones is a result of the stimulation by one hormone of the specific receptors of another.

In other species, the luteotrophic function may be performed by a different combination of hormones. In the rabbit, for example, prolactin, FSH and LH are all necessary; one function of this luteotrophic team is to stimulate the production of oestrogen which also acts on luteal cells. It is preferable, therefore, to refer to a "luteotrophic complex" rather than to Luteotrophic Hormone as being synonymous with prolactin.

Control of the corpus luteum of the sheep is somewhat different. Here, unlike the rat, the corpus luteum routinely becomes active and produces progesterone for about 15 days, in the course of a normal oestrous cycle, and then abruptly regresses (figure 4.8). If a ewe is hypophysectomized shortly after ovulation, regression of the corpus luteum occurs within 5–10 days. This can be prevented by a combination of LH and prolactin.

It seems reasonable to expect that the final regression of the corpus luteum will be preceded by a decline in these luteotrophic hormones. No such decline

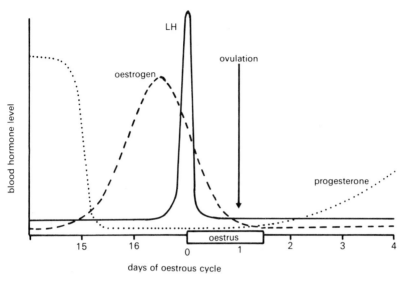

Figure 4.8 Hormonal fluctuations around the time of oestrus and ovulation in the ewe. (From Short in Austin and Short (1972)).

occurs. Instead, the sheep corpus luteum appears to be destroyed by a luteolytic stimulus, rather than by the withdrawal of pituitary support. If total hysterectomy (removal of the entire uterus) is carried out after CL formation, luteal regression does not occur at the normal time. A corpus luteum in these circumstances may persist for several months. A uterus which has been exposed for a time to progesterone evidently produces a luteolytic factor or luteolysin. Hysterectomy removes the source of the luteolysin and prolongs luteal life.

If the ovaries of the ewe are transplanted elsewhere in the body—the neck is a suitable site—the corpus luteum persists, this time for weeks rather than months. Removal of the horn of the uterus which lies on the same side as the corpus luteum also confers some prolongation of luteal life; removal of the contralateral horn of the uterus has no effect. These experiments both show that the luteolytic effect is local.

The luteolytic factor has been identified as prostaglandin $F_{2\alpha}$ (section 2.6) and its origin as the endometrium of the uterus. Injection of prostaglandin $F_{2\alpha}$ prematurely terminates the life of a CL, while countering the prostaglandin by active immunization results in prolonged luteal activity. As more than 90% of prostaglandin in the circulation is destroyed by a single passage

E

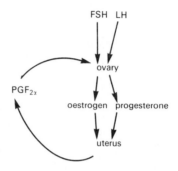

Figure 4.9   Factors controlling the oestrous cycle of the ewe.

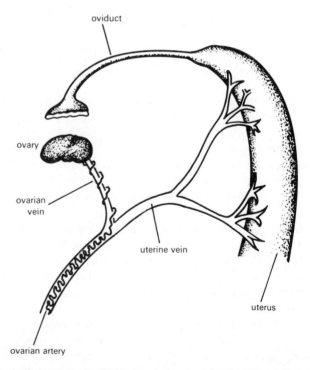

Figure 4.10   Probable route by which luteolytic prostaglandin $F_{2\alpha}$ synthesized in the endometrium of a progesterone-exposed uterus gains access to the ovary in the sheep.

through the lungs, the local nature of the luteolytic effect is explained. The main factors in the control of the sheep CL are summarized in figure 4.9.

One enigmatic feature of the action of $PGF_{2\alpha}$ is its means of access to the CL. No suitable direct anatomical connection exists. The most probable route is shown in the figure 4.10. Blood is drained from the uterus by a uterine vein. As the luteal phase approaches its end, much $PGF_{2\alpha}$ is present in this blood. Arterial blood is conveyed to the ovary (hence to the corpus luteum) by an ovarian artery which lies closely apposed to the uterine vein for part of its length. Separation of the ovarian artery from the uterine vein prevents luteolysis, and infusion of $PGF_{2\alpha}$ into the uterine vein, in amounts comparable to those present during normal luteolysis, is effective in inducing luteal regression. Probably, therefore, transfer of prostaglandin from vein to artery is by a countercurrent diffusion system. Luteolysis then occurs, either by direct action of the prostaglandin on luteal cells, or indirectly through restriction of the blood flow through the corpus luteum.

The human menstrual cycle, like the oestrous cycle of the sheep, has a routine luteal phase which lasts, on average, 14 days (figure 4.11). LH has some luteotrophic effect, and prolactin may also be involved; after total hysterectomy, however, normal hormone cycles seem to persist, and the lifespan of the corpus luteum is not extended. Injection of prostaglandin $F_{2\alpha}$ does not induce premature luteal regression and menstruation. It does not appear that the human CL, like that of the sheep, is caused to regress at the end of a cycle by the release of endometrial prostaglandin.

Another respect in which sheep and human cycles differ is that in the human a distinct follicular phase, of 16 days or so, occurs between menstruation and the next ovulation. In the sheep, only three days elapse between the fall in progesterone which marks the end of one luteal phase and the ovulation which precedes the next. Three days is insufficient time for complete development of an undeveloped oocyte into a preovulatory Graafian follicle. Clearly in the sheep a follicle which ovulates in the course of one oestrous cycle started to develop during the previous cycle. In humans, this extensive overlap between cycles will not occur. In the rat, with a virtually non-existent luteal phase, the overlap between generations of oocytes must be greater, and the oocytes ovulated in one cycle must have started their development several cycles previously.

4.4 Pituitary and brain

As in the male, the output of pituitary gonadotrophins is under the control of releasing factor from the hypothalamus. In the male, LRF production is

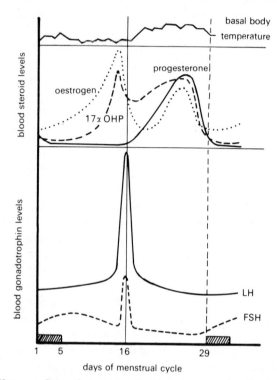

Figure 4.11   Hormone fluctuations during the human menstrual cycle. (From Short in Austin and Short (1972)).

regulated by negative feedback of gonadal steroids to the brain (section 2.7). A similar mechanism obtains in the female, although the steroids in question are predominantly oestrogen and progesterone rather than androgens. The sharp peaks of oestrogen, LH and FSH which occur prior to ovulation are the most obvious way in which the female hormonal levels differ from those of males. Such sharp rises are more characteristic of positive than of negative feedback.

The evidence for negative feedback control in the female is similar to that obtained in the male. Removal of the ovaries from female rats causes an increase in plasma levels of LH, FSH and prolactin. For some reason this rise is more sluggish than the corresponding post-castration rise in the male. Taking into account the half-life of hormone molecules and the volume within the body in which secreted hormone is distributed, it is possible to estimate actual rates of secretion. Three months after ovariectomy, the rate of LH secretion calculated in one experiment rose from approximately 1·33 μg/day to 28·8μg/day, a 22-fold increase. (The latter value represents an output of approximately 8×10^9 molecules per second.) Rises in FSH and prolactin secretion rates were less dramatic; FSH from 37μg/day to 221μg/day, and prolactin from 22 to 26 μg/day. As pituitary content of these hormones did not fall after ovariectomy, both synthesis and release must have risen, and presumably both are normally inhibited by ovarian steroids.

The rise in circulating LH and FSH which follows ovariectomy can be reversed by injecting oestrogen; this effect is detectable within one day and maximal after 2–3 days (figure 4.12). Progesterone too will reduce circulating levels of LH (but not of FSH), although only if large amounts are administered. This suggests that the smaller amounts of progesterone normally present may act synergistically with oestrogen in suppressing LH, while the failure of progesterone to suppress FSH may play some part in differential control of the two gonadotrophins. The negative feedback control of gonadotrophins in the female thus resembles that in the male, possibly with modulation by progesterone of the more significant effects of oestrogen paralleling modulation of the effects of testosterone by oestrogen in the male (see section 2.7).

A single injection of oestrogen lowers plasma LH levels in an ovariectomized rat. If, on the second or third day after the injection, a second injection of either oestrogen or progesterone is given, the result is not further depression of LH levels, but a sharp rise reminiscent of the preovulatory LH surge. In an intact animal, the preovulatory surge can be evoked prematurely by oestrogen injection. Conversely, blocking the oestrogen surge with anti-

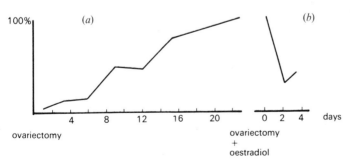

Figure 4.12   Concentrations of serum LH in female rats following (a) ovariectomy and (b) a single injection of 5μg oestradiol benzoate. To facilitate comparison, the results have in each case been expressed as a percentage of the maximum recorded in each experiment. Compare with figure 2.16. (From Gay and Midgley (1969) and McCann in Greep and Astwood (1974)).

oestrogen antibody eliminates the LH peak. Anti-progesterone antibody has no effect.

The artificial injection of oestrogen and subsequent LH rise mimic the course of events in a normal oestrous cycle, where a surge of oestrogen is followed by a rapid rise in LH and a nearly synchronous peak. This sequence is a feature of the oestrous or menstrual cycle in every species which has been studied.

In the rat, a circadian rhythm affects the response of the female to oestrogen. For a specified genetic strain of rat, and under a standardized day/night regime, the rise in LH will start at a predictable time of day; for example, in one set of experiments this "critical" period was found to lie between 2.00 and 4.00 pm. A slight rise in LH is detectable in the afternoons, even of days on which a full preovulatory surge does not take place. Treatment with barbiturates over the critical period blocks the LH rise, which demonstrates that neural events are essential for the ovulatory surges to occur.

Both positive and negative actions of steroids on LH secretion could affect synthesis or release of LRF, or both. Alternatively, steroids could act on the pituitary, altering its sensitivity to LRF supplied by the hypothalamus at a relatively invariable rate. A combination of techniques has been used to elucidate the mechanisms involved in the control of LH secretion, and hence of ovulation.

Surgical separation of the pituitary from the hypothalamus causes gonadotrophin secretion to cease. Because of this supportive function of the part of the hypothalamus overlying the pituitary—the Median Eminence— it is known as the hypophysiotrophic area. It is also possible to sever connections between the hypophysiotrophic area and the rest of the brain. If this is done, LH secretion continues, and therefore LRF is presumably still being produced and passed to the anterior pituitary. Ovulation, however, does not occur. The animals enter a permanent state of oestrus, and their ovaries soon accumulate considerable numbers of immature follicles. Apparently an isolated hypophysiotrophic area can maintain the production of constant low levels of LH but cannot respond to an oestrogen rise by stimulating an ovulatory surge of LH.

Oestrogen receptors are present in the hypophysiotrophic area; local implantation of oestrogen depresses LH output by the pituitary and a post-ovariectomy rise in LH does still occur. Progesterone implants also inhibit LH production, despite the apparent absence of specific receptors for progesterone. These findings suggest that this area of the hypothalamus receives the negative feedback effects of steroids.

Situated above the optic chiasma (the point at which many of the nerves from the retinae cross from the one side of the brain to the other) is an area known as the preoptic area (POA). Specific oestrogen receptors are present there, while electrical stimulation of the POA, or implantation of oestrogen, results in a rise in plasma LH. Behavioural oestrus and (in an immature animal) precocious puberty may also follow oestrogen stimulation of the POA (see also sections 4.5 and 8.5). The POA and the region surrounding it appear to be where positive ovulatory responses to oestrogen originate. It is not yet clear how the POA and the Median Eminence/hypophysiotrophic area interact, nor how the circadian rhythm modulates the system.

The inhibitory feedback of steroids on the ME is often known as the "long-loop" feedback. There is, in addition, a "short-loop" feedback; implants of LH in the Median Eminence suppress LH release, and the same may be true of FSH. The significance of this feedback loop is not clear, as it is incapable of preventing the LH rise after ovariectomy. Deposition of LRF in the hypothalamus may also reduce LRF output by the hypothalamus: an "ultrashort" feedback loop.

In addition, steroids act at the level of the pituitary. To investigate this, it was necessary first to eliminate hypothalamic control of the pituitary. Total suppression of endogenous LRF was achieved by tratment of rats with the CNS inhibitor phenobarbitol; it was then possible to measure variations in the response of the pituitary to a known dose of exogenous LRF. Five minutes after an injection of LRF, LH started to rise, reaching a maximum after 15–20 minutes. The size of this peak varied with the stage of the oestrous cycle, being greatest on the day of proestrus. The pituitary is most sensitive to LRF on the day in which the peak of LH would normally occur.

In further experiments, endogenous steroids were eliminated by using ovariectomized rats. In this case, the peak values of LH were lower; injection of oestradiol enhanced the response. This shows that an important variable in the control of LH release is the sensitivity of the pituitary to LRF, and that this sensitivity is determined, at least in part, by circulating oestrogen. In other experiments, oestrogen and progesterone have been found to inhibit pituitary function. Finally, the response of the pituitary to LRF is enhanced by previous exposure to LRF.

This exceedingly complex situation is summarized in figure 4.13. Despite a vast amount of information about the separate components, it is still not possible to explain why the system as a whole operates precisely as it does. Maturing follicles release oestrogen, which presumably acts on the hypothalamus to evoke LH release. In the rat, this LH rise parallels, and probably induces, a rise in progesterone, released from the interstitial tissue

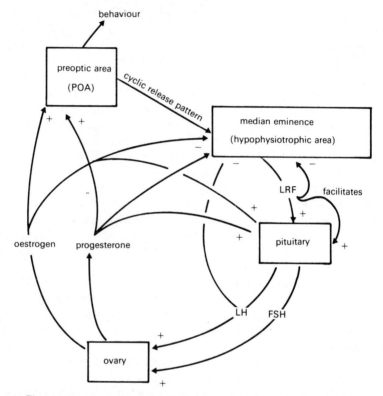

Figure 4.13 Major factors involved in the control of the rat oestrous cycle.

and not from the corpora lutea. This may facilitate further LH release, enhancing the LH peak and allowing ovulation. The initial rise in oestrogen may simply be a function of follicle maturation, in which case the maturation rate of follicles is crucial to the timing of the cycle.

Other species differ from the rat in the timing of events in the course of a reproductive cycle. The main variable is the existence in many species of a routine luteal phase. In the human, it is necessary to conclude (if only by default) that the lifespan of the corpus luteum is an intrinsic feature of the CL itself; in the sheep it is determined by the uterus. Any detailed explanation of the hormonal control of the cycles in these two species must take such factors into account.

In the rabbit, and in other species such as voles, cats and ferrets, ovulation is not spontaneous, but induced by mating. Physical stimulation of the uterine cervix is followed by a sharp rise in circulating LH, and ovulation. This can be achieved artificially by rhythmic probing of the vagina with a glass rod of appropriate dimensions. In the rabbit, a positive feedback relationship between LH and 20 α-hydroxyprogesterone appears to play a part in accelerating the ovulatory rise of LH. Even in some "spontaneous ovulators" such as man, copulation may sometimes accelerate ovulation, while in an induced ovulator such as the vole, the smell of a male can increase the normally low incidence of spontaneous ovulation. The difference between "spontaneous" and "induced" ovulators is not absolute.

Table 4.1 The main categories of reproductive cycle in mammals.

		Spontaneous ovulation	*Induced ovulation*
Corpora lutea	spontaneous	human, sheep, etc.	(unlikely to occur)
	induced by mating	rat, mouse "pseudopregnancy"	vole, ferret, rabbit, cat

Table 4.1 summarizes the main categories of reproductive cycle in mammals.

In discussing reproductive cycle, it is easy to think of the system as being a deterministic one, in which every component responds to a predictable stimulus with a precise response. This would be a remarkable mechanism. What is even more remarkable is that there is absolutely no standard pattern for oestrous or menstrual cycles. Great differences occur between individuals of a species, and great variation between successive cycles of individuals (figure 4.14). Despite this variation from cycle to cycle, most cycles in a mature

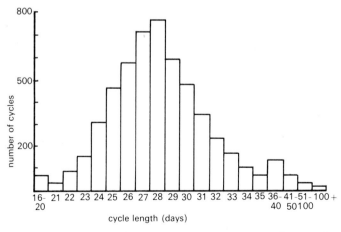

Figure 4.14   Variation in length of the human menstrual cycle, shown by a sample of 5322 cycles in 485 women. (From Parkes (1976)).

animal result in ovulation and preparation of the uterus. This emphasizes that the object of any reproductive system is to mature and ovulate eggs, prepare the uterus for possible implantation, and ensure a high probability of mating at the time of greatest fertility; not to regulate complex patterns or fluctuating patterns of hormones precisely in accordance with some preordained plan. Hormonal patterns (such as those shown in figures 4.7, 4.8 and 4.11) are merely the means of intergrating behaviour, gonads and uterus; they are not an end in themselves.

4.5 Hormones and behaviour

Behaviour, like physiology, changes in accordance with the female reproductive cycle. A female rat in oestrus is much more active than at other times, and more highly motivated towards approaching a male. If a male approaches and attempts to mate, the oestrous female adopts a characteristic posture (known as *lordosis*) with the back concave and the hindquarters raised. Lordosis is a behavioural sign of oestrus in many species, although it may vary in detail from species to species. Pig-breeders, for instance, can detect whether a sow is ready to mate by attempting to sit on her back. Only an oestrous sow remains immobile. The immobility response is more apparent if the female can smell a boar or hear his rutting call. As artificial insemination is now much used in pig breeding, boar smell is available in aerosol form.

In humans, copulation takes place throughout the menstrual cycle, and there is no parallel to the restriction of lordosis to the time of ovulation. However, behavioural and sensory changes do occur. Women show a peak in spontaneous walking around mid-cycle, while fluctuations in the sensitivity of hearing, smell and taste, in intellectual performance and in willingness to volunteer for psychological experiments, have all been described.

Ovariectomy abolishes the lordotic response, and other cyclic changes in behaviour, indicating their dependence on ovarian hormones. The loss of sexual activity after ovariectomy is very abrupt, much more so than the corresponding loss after castration of a male (section 2.7). This is hardly surprising, since in the normal cycle behaviour changes abruptly, as it must do if behavioural oestrus is to correlate closely with ovulation. In a female rat, injecting large amounts of oestrogen restores sexual motivation and lordosis. Progesterone is also effective if given to a rat previously primed with somewhat lower doses of oestrogen. The same sequence—oestrogen followed by progesterone—occurs in the course of the normal rat oestrous cycle (figure 4.7). Other rodents, in general, are similar to the rat and can also be brought into oestrus by a sequence of oestrogen followed by progesterone.

Implantation of oestrogen into the brain has been used to determine the site of its behavioural action. Lordosis is more readily evoked if the implant is in the anterior hypothalamus, particularly near the POA (see section 4.4) but some effect can be detected over quite a large area of the hypothalamus. The limbic system of the brain—above the hypothalamus and below the cerebral cortex—appears also to be involved, since damage here can affect sexual behaviour.

In contrast to the rat, behavioural oestrus in the ewe is evoked by progesterone followed by oestrogen, rather than the other way about. In this species, the ovulatory peak of oestrogen follows a fall in the high level of progesterone at the end of the preceding luteal phase (figure 4.8): here again, the artificial treatment required to induce oestrus parallels the normal sequence of hormones. As no luteal phase precedes the first cycle of the breeding season, no overt behavioural oestrus occurs in this cycle; this is known as "silent heat". In the ewe, relatively large amounts of the weak androgen androstenedione (figure 4.6) are present on the day of oestrus and may play a part in the control of oestrous behaviour.

In an induced ovulator, such as a cat, dog or rabbit, no corpus luteum is present until after ovulation. Luteal progesterone could therefore play no part in behavioural oestrus, since this inevitably takes place before mating and consequent ovulation and luteinization. In such species, oestrogen alone is sufficient to induce oestrus, and luteal progesterone may terminate it.

Androgens, rather than oestrogens, control sexual behaviour in a female monkey. The main source of these is the adrenals, not the ovaries. The absence of violent fluctuations in steroid output by the adrenals may explain why female primates are more willing to mate than most mammals at times other than midcycle. Oestrogen does play some part in regulating primate sexual behaviour, at least in the Rhesus monkey. The ovulatory surge of oestrogen stimulates secretion by the vagina of a pheromone. This communicates to the male that the female is at the peak of her receptivity and fertility. The pheromone comprises a mixture of aliphatic fatty acids; it smells of sweaty socks and has no discernible aphrodisiac effect on humans.

Female sexual behaviour is also modified, together with reproductive physiology, by external factors such as daylength. In seasonal breeders, the pineal organ modulates female reproduction in much the same way as male (section 2.7), to ensure mating, ovulation and fertilization at the most appropriate times of the year.

Reproductive cycles may be regulated also by social factors, mediated pheromonally. Female mice, grouped together in isolation from a male, show irregular cycles and prolonged periods of anoestrus. After its discoverers, this phenomenon is known as the Lee-Boot effect. Introduction of a male accelerates the onset of oestrus, and synchronizes the cycles so that the females in the group tend to come into oestrus on the same day (the Whitten effect). Even exposure to male urine causes oestrous synchrony, and the involvement of a pheromone is confirmed by abolition of the Whitten effect by removal of the olfactory bulbs of the females. A similar effect has been suggested in humans: menstrual synchrony does occur in situations where women live together in groups with some male contact, such as University residences. A phenomenon similar to the Whitten effect may underlie the advancement of puberty in female mice, and of the onset of the breeding season in ewes, by exposure to males.

One further pheromonal influence on reproduction occurs in mice. If a female has conceived by a male, and this male is then removed and replaced by a male of a different genetic strain, the first set of embryos fails to implant. The female will soon return to oestrus and mate with the alien male. This is known as the Bruce effect.

CHAPTER FIVE

EARLY DEVELOPMENT AND IMPLANTATION

5.1 Early embryonic development

After ovulation, the oocyte, still surrounded by zona pellucida and cumulus cells, is carried into the ampulla of the oviduct. Here fertilization occurs, and the zygote is transported down the oviduct, undergoing cleavage as it travels. Oogenesis is marked by growth without cell division, and results in the relatively enormous mature oocyte (100 nm); after fertilization, rapid cell division occurs without growth, and the cells of the embryo progressively revert to more typical dimensions. After a number of cell divisions, the embryo consists of a solid ball of cells, the *morula*. A fluid-filled cavity appears, and the cells become visibly differentiated into inner and outer layers (figure 5.1). This stage is known as a *blastocyst*. Entry into the uterus can occur, according to species, at any stage from one-cell (elephant shrew *Elephantulus*) to blastocyst, but morulae or early blastocysts are the more usual stages at entry.

The time of transit through the oviduct varies, but in most mammals (apart from marsupials) is 3–4 days. Retention of an intact zona appears to be necessary for embryo transport, at least in the mouse; possibly the zona

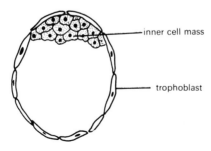

Figure 5.1 Diagram of a mammalian blastocyst

prevents adhesion to the oviduct wall, or mechanical damage. In the rabbit, an additional thick mucoid coat surrounds the zona.

Premature transfer into the uterus is fatal to early mouse and rat embryos. A sojourn in the oviduct is therefore essential before an embryo can survive the rigours of the uterine environment. Different regions of the oviduct may supply different environments; mouse zygotes cultured in isolated ampullae will develop into blastocysts, while those cultured in isthmal tissue fail to do so. Only recently has attention been directed towards the composition of the oviduct fluid and its significance as a suitable medium for embryonic development.

During the first few days of development the embryo metabolizes endogenous reserves, and consequently diminishes in mass; a cow's egg, for instance, declines by 40%. A one-cell embryo has very limited ability to take up external nutrients. Judging from the results of *in vitro* culture experiments, virtually the only utilizable energy sources are pyruvate and oxaloacetate, although glucose may be made available by conversion into pyruvate by the cumulus cells. After the first cleavage, the enzyme repertoire of the embryo increases, until by the 8-cell stage a range of carbohydrates can be used (Table 5.1). Oxygen uptake, carbon dioxide output, DNA, RNA and protein synthesis all increase markedly over the same period (figure 5.2).

Differentiation of the embryo into inner and outer cells layers is critical for development. The embryo proper derives entirely from the inner cell mass: the outer cell layer, or *trophoblast*, has a number of important functions, but does not form any part of the young animal. This has been most elegantly demonstrated by Gardner, using fine surgical techniques to separate and recombine the different components of mouse blastocysts. Trophoblast alone will implant, but fails to form any embryo; isolated inner cell mass fails to implant. Inner cell mass from one strain of mouse combined with trophoblast from another strain transferred into the uterus of a foster female gives an embryo derived solely from the donor of the inner cell mass. The primary function of the trophoblast is as a vehicle for the embryo, to ensure successful establishment and continuation of pregnancy (sections 5.3, 6.1 and 9.5).

In a species which produces many young in a litter, the embryos tend to be evenly spaced in the uterus. Spacing of the blastocysts must occur before implantation starts. In the rabbit, it has been shown that distension of the uterus by an object evokes peristaltic waves, in *both* directions, from the point of distension. This is probably the means of spacing throughout each horn of the uterus. Depending on the number of eggs ovulated, and the anatomy of the uterus, migration of blastocysts between the horns of the uterus may

Table 5.1 Energy sources which can support the *in vitro* development of mouse oocytes and embryos.

Substrate	Stage of development			
	Oocyte	1-cell	2-cell	8-cell
pyruvate	+	+	+	+
oxaloacetate	+	+	+	+
lactate	−	−	+	+
phosphoenolpyruvate	−	−	+	+
malate	−	−	+	+
citrate	?	?	−	+
α-oxoglutarate	?	?	−	+
acetate	?	?	−	±
D-glyceraldehyde	?	?	−	−
glucose-6-P	?	?	−	−
fructose-6-P	?	?	−	−
glucose	−	−	−	+
fructose	?	?	−	±

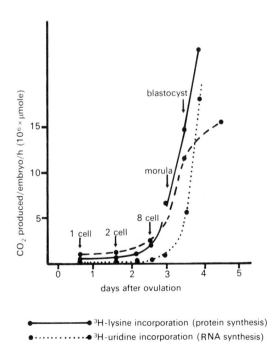

——•—— ^{3}H-lysine incorporation (protein synthesis)

•············• ^{3}H-uridine incorporation (RNA synthesis)

Figure 5.2 Metabolism of the preimplantation mouse embryo: carbon dioxide output, RNA and protein synthesis. (From McLaren in Austin and Short (1972)).

occur. Movement of blastocysts presumably ceases either when a blastocyst expands beyond a certain size, or when it actually attaches to the endometrium at the start of implantation.

In the majority of species the blastocyst spends a few days at most in the uterus before a more or less rapid implantation. The ungulates differ from this pattern. The blastocyst of the sheep extends, by fluid accumulation, into a vesicle up to 190 mm long before a lengthy process of implantation some 15–17 days after fertilization. In the cow and horse 40 and 50–60 days respectively may elapse. Until implantation, nutrients are obtained from "uterine milk", a mixture of endometrial secretions and tissue debris, and specialized cells develop on the trophoblast to facilitate uptake of this.

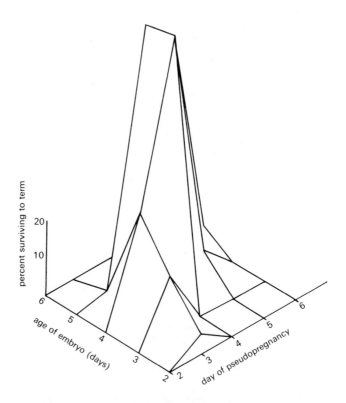

Figure 5.3   Three-dimensional diagram showing the proportion of successful implantations which resulted, following the transfer of mouse embryos of various ages to uteri at different times after ovulation. For successful implantation, synchrony of embryo and uterus is necessary. (Data in Noyes, Dickmann, Doyle and Gates in Enders (1963)).

5.2 Changes in the uterus

Development of the embryo is closely synchronized with changes in uterine structure and function. However, a rat blastocyst transferred into a uterus which is a day more advanced than itself implants at the time expected from the stage of the uterus, indicating that the uterus has a greater influence over the precise timing of events. The critical importance of timing, and specifically of synchrony of uterine and blastocyst development, is illustrated in figure 5.3.

The processes which precede reception of an implanting embryo can be divided into presensitization and sensitization, although the latter may start before presensitization is completed. During the presensitization period, the lumen—complex and branched in an oestrous mouse—becomes simple and sinuous. There is a transient rise in mitotic activity in glandular tissue and epithelium (figure 5.4), and the surface of the latter becomes corrugated and covered with regular microvilli. During the sensitization phase there is an increase in cell division in the stroma (figure 5.4), which also tends to become oedematous. The lumen of the uterus virtually disappears, with very close contact between the apposed luminal epithelia, and the flattening of the microvilli. If a blastocyst is present, it will be held closely against the luminal epthelium; implantation then follows.

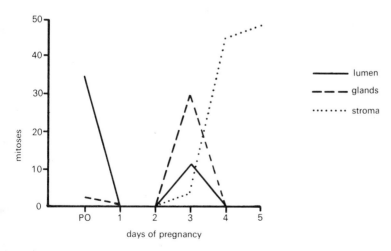

Figure 5.4 Distribution of mitoses in tissues of the mouse endometrium. Mitosis was stopped by colchicine treatment 2 hours before examination, and the number of mitotic figures in one random cross-section from each uterus counted. (From Finn and Porter (1975)).

Ovariectomy shortly after fertilization prevents these changes, showing their dependence on ovarian hormones; as presensitization and sensitization occur only during pregnancy or pseudopregnancy, progesterone is presumably involved. Ovariectomy after a delay may permit presensitization but block the changes associated with sensitization. This is because in the intact animal oestrogen is also needed for sensitization and successful implantation. The requirement for full preparation of the uterus is therefore a sequence of progesterone followed by oestrogen.

The normal pattern of endometrial mitosis can be simulated by appropriate hormone treatment after ovariectomy. Progesterone elicits the increase in cell division within the epithelial and glandular tissues, characteristic of presensitization, and oestrogen the shift to stromal mitosis which characterizes the sensitization phase. This transition is more rapid if the progesterone is preceded as well as followed by oestrogen.

The hormone treatment required artificially to duplicate in an ovariectomized animal early events in normal pseudopregnancy resembles the hormonal sequence to which the uterus of an intact animal is exposed. An ovulatory surge of oestrogen is followed by a rise in progesterone; shortly before the time at which implantation may take place there is evidence, in some species at least, of a further rise in oestrogen (e.g. figure 4.11).

Not all species require oestrogen for implantation; in some, progesterone alone is sufficient. Oestrogen may be of greater significance in species such as the mouse, where delayed implantation (section 5.4) occurs during lactation. The dependence of implantation on oestrogen may permit control to be more conveniently exerted over the timing of implantation.

5.3 Implantation: interactions between embryo and uterus

The sequence of events in implantation is more clearly established than their causal relationships. The blastocyst first sheds its zona pellucida and adheres to the uterine epithelium. At about this time, the lumen of the uterus closes completely round the blastocyst and holds it securely in place. Any further spacing of implanted embryos takes place by differential elongation of the uterus.

The embryo becomes activated, with acceleration of many metabolic processes. Carbon dioxide output may be taken as an example (figure 5.2). Activation is followed by a considerable uptake of water and consequent increase in volume by as much as 10- or 20-fold.

Uterine capillaries and the uterine wall in the immediate vicinity of the embryo become more permeable; resulting from this there is a local stromal oedama and a rise in the oxygen tension within the uterus. If a dye of high M.Wt. such as Pontamine Blue (M.Wt = 993) is injected at this time (during day 4 of pregnancy in the mouse), the increase in capillary permeability allows molecules to diffuse out and stain the surrounding tissue. The visible result is a blue zone corresponding to each implantation site, providing a useful experimental tool.

Shortly afterwards, the endometrium around the embryo shows the first signs of a Decidual Cell Reaction (DCR). This consists of the disruption of the epithelium and transformation of the loosely-packed fibroblast-like cells of the stroma into large rounded glycogen-filled cells tightly bound to each other by gap junctions. Large amounts of an alkaline phosphatase (of unknown function) characterize the decidual tissue. The DCR originates next to the blastocyst and spreads progressively until a considerable mass of endometrium is involved in the formation of an "implantation chamber" round the embryo.

After decidualization the tissue is, in effect, committed to pregnancy; the cells involved cannot revert to simple stromal cells, but will be jettisoned at the end of pregnancy with the reformation of the epithelium beneath the decidual tissue. Primate endometria show a form of decidualization, not evoked by an embryo, in the course of the menstrual cycle.

The function of decidual tissue is surprisingly poorly understood. As decidual cells contain large amounts of glycogen and are metabolically very active, it is probable that they have a nutritive role before the establishment of a functional placenta.

In mice, regression of the decidual cells begins early in pregnancy, according to a strict time scale and pattern. (This contrasts with primates, where the life of decidual cells can be prolonged by hormone treatment.) Cells bordering the lumen die first. This results in a progressive enlargement of the implantation chamber which may be important in accommodating a growing embryo. By this stage of pregnancy, the true placenta is functional, and the nutritive role of the decidua probably superfluous.

Decidualization of the uterus is accompanied by striking changes in the blastocyst. The trophoblast cells enlarge, divide rapidly, and start to invade the decidual tissue. As they do so, differentiation may take place into cellular and syncytial regions. Some highly invasive giant cells also form. Trophoblast invasion is particularly obvious in the area which overlies the inner cell mass; an inductive stimulus of inner cell mass on trophoblast appears to be involved. This leads to the formation of a distinct "ectoplacental cone" of trophoblast over the embryonic region of the mouse conceptus.

Depending on species, the expanding trophoblast may destroy, pass between, or fuse with epithelial cells. It then advances through the decidua, destroying and phagocytozing decidual cells. Eventually, after close contact has been made with uterine blood vessels, growth of trophoblast as a tissue virtually ceases, although continued division may result in formation of further cells. In some species, large numbers of trophoblast cells are deported from the uterus in the blood circulation, and settle in the lungs or some other suitable part of the mother's body.

Uncontrolled growth of trophoblast would clearly be damaging, and possibly lethal. The decidua appears to play an important part in the control of trophoblast invasion. Trophoblast transplanted to the undecidualized uterus of a mouse during an oestrous cycle penetrates deeply into the myometrium. Implantation into a pseudopregnant uterus, on the other hand, is followed by a DCR and invasion restricted to the endometrial mucosa. A further measure of control may be inherent limitation of the lifespan of trophoblast. Mouse trophoblast transplanted ectopically (that is, to a position other than the normal one) dies at a time corresponding approximately to full term of the conceptus from which it was obtained.

As ever, the mouse and rat supply most of the information on the nature of implantation, but cannot be assumed to be typical of all mammals. In some species, the DCR is absent, and uterine response to the presence of an embryo minimal (e.g. ferret).

The function of trophoblast invasion is therefore to erode the uterine lining, so bringing fetal and maternal tissues into closer proximity, leading to the formation of the definitive placenta, and facilitation of the mutual transfer of materials. Trophoblast corresponds to the chorionic element of the mature placenta, and differences in the pattern of trophoblast invasion are reflected in different placental structures (section 6.2). Some of the major patterns of implantations are shown in figure 5.5.

Implantation results not from parallel and independent changes in embryo and uterus, but from complex interactions between the two. McLaren aptly compared those to the courtship behaviour of the stickleback to illustrate the point: an action of one animal releases behaviour of the other, which in turn acts as a stimulus on the first animal, and so on. These reciprocal interactions, with each step conditional on a preceding one, culminate in mating. Analogous interactions occur between embryo and uterus.

The earliest signs of implantation apparent in the mouse blastocyst—loss of the zona and adhesion to the epithelial surface—are due to an action by the uterus. An oestrogen-treated uterus produces a proteolytic enzyme, present in greatest amount on day 3 of pregnancy. This lyses the zona and so permits adhesion of the embryo to the uterine epithelium.

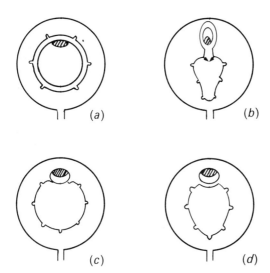

Figure 5.5 Forms of implantation found in mammals. (*a*) Centric (*b*) Eccentric (becoming secondarily interstitial) (*c*) partly interstitial (*d*) interstitial. (From Wimsatt (1975)).

Activation of the blastocyst has also been attributed to a further direct action of oestrogen, but the evidence for this is not very good. From experiments *in vitro*, glucose appears to be a critical requirement for further development of blastocysts. In a medium lacking glucose, blastocysts remain in a quiescent state. There is no direct evidence for restraint of a blastocyst by nutrient restrictions *in vivo*; however, it is possible that the increase in permeability of the uterine wall allows increased amounts of glucose (or other nutrients) to reach the embryo. A rise in pO_2 in the lumen also results from the permeability increase, and might also stimulate the blastocyst. There is some evidence for an inhibitor present in uteri with resting blastocysts. In this case, activation would result from withdrawal of inhibitor rather than presence of a stimulus.

In the rabbit and a few other species, a specific stimulus emanating from a progesterone-treated uterus has been identified, necessary for full expansion of the blastocyst (although expansion can be obtained *in vitro* without it). This is a protein, of molecular weight 15,000 named "blastokinin" and "uteroglobin" by the two research groups which independently established its existence. Blastokinin is found in the uterus only at the time of implantation, when it may constitute as much as 40% of total protein in the uterine fluid.

The DCR provoked by an embryo in some ways resembles the formation of a granuloma around a foreign body elsewhere in the body, and may represent a highly specialized form of granuloma response. It is therefore not surprising that it can be elicited by a variety of stimuli other than the normal one. Among the most effective is the intraluminal injection of a small droplet of oil. The precise nature of the stimulus presented by the oil is not clear.

Injection of a small bubble of air also elicites a DCR. If CO_2 is removed from the air beforehand, the DCR is much reduced; increasing the CO_2 content enhances the DCR. An obvious conclusion to draw is that activation of a blastocyst greatly increases its CO_2 output (figure 5.2) and that this is a major component in evocation of the DCR.

Pig and rabbit blastocysts can synthesize steroids, and rabbit blastocysts carry significant quantities of oestrogen. Intraluminal injection of an oestrogen antagonist into one horn of the uterus of a rabbit on day 5 of pregnancy inhibited implantation. This suggests that the surface-bound oestrogen may be important in implantation, possibly acting as a stimulus to the uterus, and that in the rabbit oestrogen *is* needed for implantation (cf. section 5.2) albeit locally applied in the uterus and not in the maternal circulation.

Blastokinin has some progesterone-binding properties (although different research groups differ in their assessments of its affinity for progesterone) and it is possible that the function of blastokinin is to convey progesterone to the blastocyst for conversion to oestrogen (see figure 4.6). A gonadotrophin resembling LH has also been detected on the rabbit blastocyst and may be involved in the control of embryonic steroidogenesis.

Further actions of the uterus on the blastocyst may be required before pregnancy is established. If a blastocyst is detained artificially in the oviduct it may subsequently form trophoblast and implant, but embryonic tissue will not differentiate. Evidently a uterine environment is necessary. Blastocysts removed from the uterus and cultured in a standard medium maintain a high level of metabolic activity, but do not show trophoblast outgrowth. This will, however, occur if small amounts of fetal calf serum are added to the medium. Some macromolecular component must therefore be supplied by the uterus for normal trophoblast outgrowth.

Treatment of a female with actinomycin D (an inhibitor of transcription of nuclear DNA) prevents the degeneration of the luminal epithelium which normally accompanies trophoblast invasion. This suggests that destruction of epithelial walls is a genetically programmed function of the cells themselves. The action of invading trophoblast may therefore be inductive rather than destructive. Inherently programmed cell death is a common occurrence in development, not least in the eventual death of trophoblast itself.

Some of the postulated interactions between blastocyst and the uterus are summarized in figure 5.6.

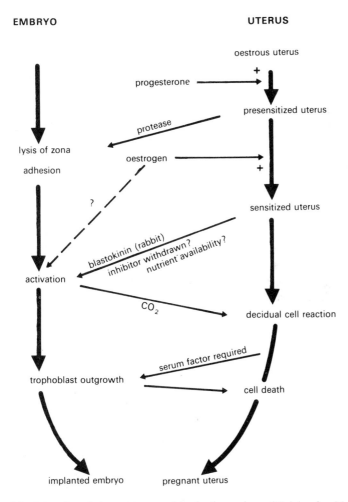

Figure 5.6 Interactions between uterus and implanting embryo. (Mainly after McLaren (1969)).

5.4 Delayed implantation

In some species, implantation does not immediately follow development of the embryo into a blastocyst. Indeed, it may be delayed for a considerable period of time. Delayed implantation can be either obligate or facultative. In the former category are animals such as badgers, stoats, some deer and seals, and certain bats, in which a delay in gestation makes it possible for both mating and birth to occur at convenient times of year, irrespective of the length of the gestation period. The European badger (*Meles meles*), for instance, mates in February, implants in December and gives birth early in the following year; the roe deer (*Capreolus capreolus*) mates in July or August, implants in January and gives birth in May; and the grey seal (*Halichoerus grypus*) mates in autumn, implants in April and gives birth in July or August.

Facultative delays in implantation are found in the mouse and rat. These species show a postpartum oestrus, as a result of which a second conception may occur immediately on completion of the first pregnancy. The metabolic strain on the female of coping simultaneously with two litters—one in the uterus, the other suckling—is lessened by delaying the implantation of the second litter until lactation has declined. In the rat, the interval between postpartum mating and birth of the second litter may be 35 days, compared with a normal gestation period of 22 days.

Metabolic activity of a quiescent blastocyst is much lower than normal; CO_2 output and protein synthesis are greatly lowered, and DNA synthesis negligible. The causes of this depression in metabolism are complex, probably differ from species to species, and cannot be fully elucidated until the control of normal implantation is clarified. Lack of a uterine activating stimulus is an obvious possibility. In the roe deer (*Capreolus capreolus*), an obligate delay appears to be due to the lack of certain essential factors supplied by endometrial gland secretions as development recommences. The nature of the limiting factors among the components of the secretion is not known; proteins, amino acids, glucose, galactose and fructose are all present in the secretion, and all might stimulate blastocyst development.

Nutrient limitation may also affect facultative implantation delay in the rat. This is suggested by the fact that the nutritional state of the female has a direct influence on the length of delay. Greater delays occur with larger litters, and restricting the food intake of the lactating female also extends the delay. If, on the other hand, embryonic diapause is controlled by the presence of an inhibitor rather than the lack of a positive stimulus, the immediate control of delayed implantation would involve the withdrawal of inhibition.

Whatever mechanism immediately limits blastocyst development, resumption of development in many species is conditional on the presence of oestrogen. During lactational delay of implantation in the rat, the uterus has completed presensitization but has not embarked on the final process of sensitization. This indicates that oestrogen is lacking, and suggests that failure to implant during lactation, and after ovariectomy (section 5.3) are similarly caused. Oestrogen injection during delay results in sensitization of the uterus, resumption of embryonic development, and implantation.

In other species, embryonic diapause is controlled otherwise than by withholding oestrogen. In the European badger (*Meles meles*) low prolactin levels have been implicated, while in the armadillo (*Dasypus novemcinctus*) ovariectomy during delay brings about implantation. Ovarian hormones—probably progesterone—seem positively to inhibit implantation. In the bat *Artibeus* the quiescent phase of the embryo occurs *after* implantation. In these species, different mechanisms must have evolved to achieve the advantages of embryonic diapause.

CHAPTER SIX

PREGNANCY

6.1 Hormonal factors in pregnancy

6.1.1. *Maternal recognition of pregnancy*

Gestation depends on the persistence of a decidualized uterus, and therefore on the continued presence of progesterone. The initial source of progesterone is the corpus luteum. In eutherian mammals (but not in most marsupials) pregnancy lasts longer than the period of luteal activity in an oestrous cycle. Fertilization must in some way rescue the corpus luteum from its fate and cause it, instead of declining, to maintain progesterone output in support of a pregnant uterus, and to suppress ovulatory LH surges for the duration of pregnancy. The "decision" of the corpus luteum to persist rather than degenerate is a crucial aspect of the maternal recognition of pregnancy.

In a few species, such as the ferret, copulation induces a pseudopregnant state which lasts as long as the gestation period. Here the mother recognizes mating, not fertilization, and no maternal recognition of pregnancy, as such, is necessary. Following an infertile mating the female is pseudopregnant, and therefore unable to ovulate and conceive, for the same time as if she were actually pregnant. In the rat, the pseudopregnancy is shorter than pregnancy, but long enough to allow the embryo to reach the uterus, implant, and signal its presence back to the ovary. If it were not for pseudopregnancy, the corpora lutea of the oestrous cycle would have regressed totally several days before implantation. If conception does not follow mating, the female will have another opportunity when she returns to oestrus after twelve or so days. If mating itself does not occur, no full luteal phase intervenes and the return to oestrus is more rapid—four or five days. Other species (e.g. sheep, human) have a full luteal phase in the course of every cycle, shorter than pregnancy but long enough, as in the pseudopregnant rat, for the conceptus to signal its existence.

Three means of communication from conceptus to the corpus luteum are possible. Neural information from a pregnant uterus might reach the central nervous system and, by way of the hypothalamus, stimulate and maintain the production of pituitary luteotrophic hormones. Secondly, the conceptus might produce its own luteotrophin, acting directly on the corpus luteum to maintain the production of progesterone. Finally, in species in which the lifespan of the corpus luteum is determined by the luteolytic influence of the uterus (section 4.3), the presence of a conceptus within the uterus must prevent the production of luteolysin. All three mechanisms are important, although to different extents in different species. The endocrine basis of pregnancy is best explained by considering the situation in a variety of species.

6.1.2 *Rat*

In the course of an infertile cycle, rat corpora lutea never become fully active, progesterone output remains low, and oestrus soon recurs. Whether or not mating achieves fertilization, it will probably induce pseudopregnancy (section 4.4); if this does happen, the transient corpora lutea become fully active and maintain a high level of progesterone for 10–12 days (figure 6.1). Enhanced and prolonged luteal activity is presumably due to a large extent to the rise in both LH and prolactin which follows copulation. The neural stimulus of copulation has led, through afferent neural pathways, to an increase in output of luteotrophic hormones.

Pseudopregnancy by itself does not cause the maintenance of CL and uterus for the full gestation period of 23 days. Further stimulation is required. If decidualization is induced artificially (section 5.3), further extension of the

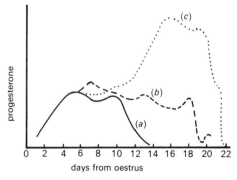

Figure 6.1 Progesterone levels in the female rat during (*a*) pseudopregnancy, (*b*) pseudopregnancy and the induction of a Decidual Cell Response, and (*c*) pregnancy. (From Pepe and Rothchild (1974)).

luteal lifespan occurs (figure 6.1). Progesterone levels similar to those of pseudopregnancy persist until about the 18th day after mating or its simulation. Decidualization is also recognized by corpora lutea in the ovary; the mechanism is not known.

Even a decidualized uterus does not cause progesterone secretion to rise to the levels found in pregnancy. A further luteotrophic stimulus must occur at around day 10–12 of pregnancy, since it is then that progesterone levels of pregnancy diverge from those of pseudopregnancy and decidualization (figure 6.1). This stimulus cannot be pituitary LH or prolactin, since these do not rise at mid-term, and it must be contingent on the presence of an actual conceptus, not on the occurrence of mating or decidualization.

Such a luteotrophin has been identified in the rat placenta; it appears to be a protein of molecular weight between 25,000 and 50,000, with a trophic effect on mammary gland tissue as well as on the corpus luteum. It has therefore been designated Rat Chorionic Mammoluteotrophin (RCM) and is, gratifyingly, present in greatest concentrations on the 12th day of pregnancy. It probably originates in the syncytiotrophoblast. Curiously, trophoblast or entire blastocysts transplanted to ectopic sites invade their new sites enthusiastically but do not interrupt the sequence of oestrous cycles. Some interaction with decidual tissue is apparently necessary, and this is lacking in an ectopic site.

Pregnancy in the rat thus falls into two main phases. Post-coital rises in LH and prolactin inform the CL that mating has occurred and induce full luteal activity and progestational changes in the uterus. Hypophysectomy, or treatment with LH antiserum, is followed by luteal and uterine regression, reversible by injection of LH or progesterone, the latter effect being enhanced by oestrogen. Ovariectomy likewise causes the uterus to regress. If fertilization, decidualization and implantation occur, the decidua and conceptus exert on the uterus a luteotrophic influence which is appreciable by day 10–12 of pregnancy. Uterine regression is no longer the sequel to hypophysectomy.

Late in pregnancy, ovariectomy no longer terminates pregnancy. If ovarian steroids are not essential, steroids from some other source may be important. Adrenal progesterone does contribute, but the steroids of late pregnancy are produced largely by the placenta, and applied directly to the endometrium. Placental steroid synthesis, if it is under external control, may be influenced either by the placental luteotrophin or, possibly, by the rise in pituitary LH and prolactin which follows the fall in circulating progesterone levels near term.

The conceptus therefore achieves progressively greater endocrine autonomy, and a correspondingly increased measure of control of its own destiny (figure 6.2). Eventually neither pituitary, ovary, nor indeed fetus are necessary for pregnancy to continue to term.

6.1.3 Human

The human blastocyst implants on the sixth day after fertilization. In an infertile cycle, the corpus luteum begins to regress on about the twelfth day. Maternal recognition of pregnancy must occur between these two days. By day 10, a new hormone is detectable in blood. This is a glycoprotein (M.Wt. 30,000) secreted by trophoblast and known from its site of origin and principal biological effect, as Human Chorionic Gonadotrophin (HCG). Because it is unique to pregnancy, this is the basis of most methods of pregnancy diagnosis.

Although it can in some circumstances stimulate follicle growth (cf. FSH, section 4.3), HCG chiefly resembles LH in its properties. (An interesting difference is in the half-life of HCG molecules: 40 hours, compared with less than 30 minutes for LH.) Injection of HCG during the luteal phase of a menstrual cycle stimulates the activity of the corpus luteum and prolongs its life; the most probable explanation for its role in pregnancy is therefore that it is a luteotrophin. However, in some cases when HCG injection has been

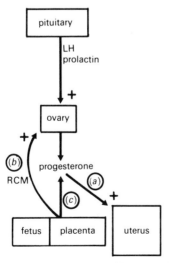

Figure 6.2 Hormonal control of gestation in the rat (*a*), (*b*), and (*c*) refer to early, middle and late pregnancy.

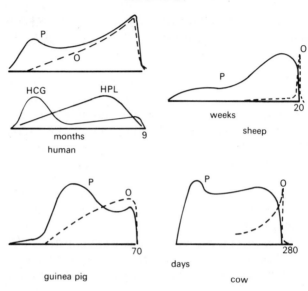

Figure 6.3 Hormone levels during pregnancy in human, sheep, guinea pig and cow. P = progesterone, O = oestrogen (From Bedford, Challis, Harrison and Heap in Perry (1972))

followed by observation of the ovaries, two corpora lutea have been found. This suggests an alternative possibility, that HCG might operate by inducing ovulation and a second corpus luteum to replace or supplement the first.

Figure 6.3 shows the serum levels of progesterone and HCG throughout pregnancy. Serum progesterone increases sharply as HCG rises to a peak, further evidence of HCG's luteotrophic role. Thereafter HCG remains at a low level until the end of pregnancy, with a slight rise to the 30th week. Progesterone, on the other hand, increases progressively until shortly before birth. This second rise in progesterone (and a similar rise in oestrogen) cannot be dependent to any extent on HCG; nor can it be luteal, since ovariectomy after the 24th day of pregnancy no longer causes abortion. As a source of steroids, the corpus luteum is progressively replaced by the placenta.

Endocrine relations between mother, placenta and fetus are complex, with steroid hormones and their precursors and products being passed to and fro. The placenta obtains cholesterol from both mother and fetus, and returns it as progesterone (figures 4.6 and 6.4). In the fetal adrenals, progesterone is converted to dehydroepiandrosterone (DHEA), as well as corticosteroids such as cortisol. Conjugated with sulphate, DHEA passes back to the placenta, where it may be converted to oestrone and oestriol. Some oestrone returns to the fetal liver; here (together with rather larger amounts of the conjugated DHEA) it is hydroxylated before returning yet again to the placenta. The hydroxylated steroids are finally converted to oestriol, which is ultimately secreted in the urine (figure 6.4). The complementarity of steroidogenesis in placenta and fetus has led to the concept of the "feto-placental unit".

The concentration of a hormone in the circulation is determined by its rate of clearance as well as by its rate of synthesis. In women, the rate of clearance does not alter significantly during pregnancy, and the increase in plasma levels is largely the result of an increase in the rate of synthesis, which may reach 300 mg/day. (In other species, such as the guinea pig, in which pregnancy similarly depends on placental progesterone, a rise is achieved by a rapid increase in concentration of a serum protein with a high affinity for progesterone, and consequent lowering of the rate of clearance of progesterone from the circulation.) The rate of secretion of steroids by the human placenta is closely related to the size of the placenta. It is possible, therefore, that the most significant

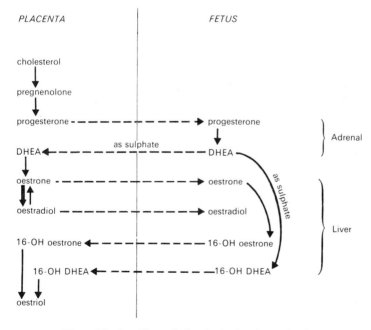

Figure 6.4 Steroid metabolism in the fetoplacental unit.

factor in the control of total steroid output is the rate of placental growth. However, HCG has been found to stimulate steroid synthesis *in vitro*, including the aromatization required for oestradiol synthesis and the conversion of oestradiol to oestriol. It may therefore play a significant part in the control of placental, as well as luteal, steroid secretion.

The placenta produces a further protein hormone, human placental lactogen (HPL), also known as human chorionic somatomammotrophin (HCS) (figure 6.3). As its name suggests, it has a variety of effects. Like pituitary growth hormone, which it closely resembles in structure, it has growth-promoting properties; the fact that its potency is much less than that of GH may be offset by the rapid rate of synthesis, of the order of 1 g/day. HPL is also lactogenic, resembling prolactin in this respect (see section 7.4), and has an assortment of metabolic effects on the mother during pregnancy (section 6.3).

The similarity between HPL and prolactin probably explains why the former is luteotrophic in the rat. Although prolactin has little influence over the human CL, HPL may well act synergistically with HCG to control steroid production in CL or placenta.

The human placenta is probably one of the sources of yet another protein hormone, relaxin. This is found during pregnancy in a number of species, and allows relaxation of the pubic symphysis and dilation of the uterine cervix, facilitating the departure of the fetus at birth. As concentrations of relaxin increase steadily throughout pregnancy, a role in accommodation of the growing fetus is also possible. Relaxin is also produced in the corpus luteum of pregnancy.

The main hormonal factors in human pregnancy are shown in figure 6.5.

6.1.4 *Horse*

In the mare, unfertilized eggs are retained within the oviduct, and only fertilized ones are permitted to enter the uterus. Pregnancy is therefore recognized extremely early, although the implications of the ability to discriminate between fertilized and unfertilized eggs are not clear.

The horse embryo passes into the uterus where, still as a blastocyst, it may grow to a size of 6–7 cm. Embryonic development proceeds, and the extraembryonic membranes—chorion, allantois, yolk sac—form. From about the 25th day after conception, rapid growth occurs among the cells which lie at the junction between yolk sac and allantois. A thickened band of tissue develops around the embryo; this is referred to as the chorionic girdle. Closer examination reveals the girdle to be composed of numerous villi of

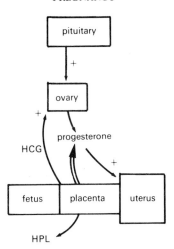

Figure 6.5 Hormonal control of human gestation.

trophoblast cells. These grow out, penetrate the endometrium, eventually (by day 37–38) lose contact with the conceptus and thereafter remain separate from it. While the conceptus itself is not yet closely associated with the endometrium, a considerable area of the endometrium has therefore been infiltrated by cells of conceptal origin.

Regression of the centre of the area of infiltration and outgrowth of its periphery results in the formation of "endometrial cups" containing numerous cells of conceptal origin. The definitive implantation then takes place, in these endometrial cups, from about day 45–50 of pregnancy.

The function of the endometrial cups appears to be endocrine. A hormone, Pregnant Mare Serum Gonadotrophin (PMSG) is secreted by the infiltrated embryonic cells, and is present in the maternal circulation between formation and regression of the cups—approximately days 40–150 of pregnancy. During this period occur a number of secondary ovulations, together with luteinization of immature follicles. This supplements the primary corpus luteum, which starts regression at day 12. The effect of PMSG is therefore luteotrophic, maintaining progesterone output during the first half of pregnancy.

After day 150, the accessory corpora lutea also regress and serum progesterone levels fall, before sharply rising again during the last few weeks of pregnancy (which lasts 336 days). The source of this final rise in progesterone is the placenta, which is very active also in oestrogen secretion.

6.1.5 *Sheep*

In the sheep oestrous cycle, the lifespan of the corpus luteum is determined not by fluctuations in luteotrophin levels, but by the secretion of an endometrial luteolysin, identified as prostaglandin $F_{2\alpha}$ (section 4.3). The presence of an embryo in the uterus must block the production of this luteolysin. The antiluteolytic mechanism is unknown, but it appears to be chemical, since luteolysis can be postponed in the sheep by intrauterine infusion of a macerate of tissue from sheep embryos. A similar macerate of pig embryonic tissue is ineffective. Subsequently, the conceptus becomes luteotrophic as well as antiluteolytic. Placental luteotrophins and steroids are involved in much the same ways as in other species.

6.1.6 *Rabbit*

The first signs of pregnancy in the rabbit are detectable as early as day 5; a pregnant female has circulating progesterone levels more than twice as high as a female mated with a vasectomized male. The significance of this is not known.

In contrast to the species previously discussed, ovariectomy or hypophysectomy of a pregnant rabbit, at any stage of pregnancy, are both followed by abortion. The placenta is apparently incapable of producing luteotrophin to maintain the corpora lutea, or placental progesterone as a substitute for luteal. The corpora lutea, throughout pregnancy, are controlled by a luteotrophic complex of prolactin, FSH, possibly LH and follicular oestrogen.

6.1.7 *Some eccentric forms*

In a pregnant elephant, progesterone is virtually undetectable in the corpora lutea, and appears to be totally absent from the circulation during pregnancy. Either it is not required for maintenance of the uterus, or the uterus is more sensitive to progesterone than current methods of assay. A further unexpected feature of elephant reproduction is that up to 40 corpora lutea may be present in the ovary; perhaps multiple ovulation and the accumulation of a certain mass of luteal tissue must precede pregnancy.

Another anomalous species is the plains vizcacha (*Lagostomus maximus*), a South American hystricomorph rodent. This animal ovulates up to 800 eggs, and forms a very large number of corpora lutea. Most eggs are not fertilized and degenerate; only 7 implant, and of these only two survive until birth. Again, a large number of accessory corpora lutea are required to maintain pregnancy.

6.1.8 *Endocrine control of gestation*

Eutherian mammals therefore vary greatly in the hormonal control of their gestation. In all cases (with the possible exception of elephants) there is a period when post-ovulatory corpora lutea produce progesterone to maintain the uterus. The period of luteal activity is generally extended, compared with that of the normal oestrous cycle. In mice and rats, prolongation and pseudopregnancy are induced by the physical act of copulation; in the ferret, pseudopregnancy lasts as long as pregnancy, and the female is indifferent to the presence or absence of a fetus. In others, prolongation of luteal life is by placental gonadotrophins; these may also (as in the horse) induce a relay of secondary corpora lutea. Finally, in most species the placenta produces steroids and maintains the uterus independently of the corpora lutea. The timing and emphasis of pituitary gonadotrophins, placental gonadotrophins, and placental steroids in the control of pregnancy differs from species to species, but the same themes recur.

6.2 Structure of the chorio-allantoic placenta

Placentae differ considerably in their gross structure, although not according to any clear phylogenetic pattern. In the horse, pig, camel, whale and certain other mammals, the villi which interdigitate with the endometrial surface are spread over virtually the whole surface of the chorion. This is known as a *diffuse placenta*. In other species, chorionic villi are localized in their distribution. Cats and dogs have zonary placentae, with chorionic villi restricted to an equatorial band; in the mink, the placental region is a simple patch. Ruminants generally have placentae in which the villi are in numerous small cotyledons; in the sheep, there are between 60 and 100 of these. Finally, in humans, rats and mice, the placenta is discoid and the chorionic villi organized into a circular plate (figure 6.6).

Placentae are diverse also in fine structure. Most classifications of placental fine structure are based on the number of tissue layers intervening between maternal and fetal circulations, and derive from a scheme originally proposed by Grosser. According to Grosser's classification, there are three relevant tissue layers on the maternal side, and three on the fetal side of the placenta: endothelium around the uterine capillaries,

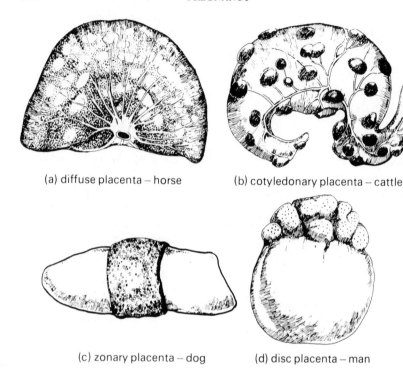

(a) diffuse placenta – horse   (b) cotyledonary placenta – cattle

(c) zonary placenta – dog   (d) disc placenta – man

Figure 6.6   Forms of mammalian placenta. (From Witschi, E. (1956), *Development of the Vertebrates* (W. B. Saunders) (fig. 247, p. 378).

uterine epithelium, trophoblast (chorion), fetal connective tissue, and fetal endothelium. In the pig and horse placentae, all six layers are present.

Other forms of placentae can be derived from this arrangement by the loss of one or more of the intervening layers. Thus the placenta of the sheep has lost the maternal epithelium, and so on. The full range of types, according to the Grosser classification, is given in Table 6.1.

This scheme was thought to reflect important physiological differences in placental "efficiency"; the fewer the interposed layers, the more readily would materials cross the placental barrier. In this case, it is surprising to find that species with the "inefficient" six-layered placenta have young which may be very well developed at birth, the horse being a good example of this. As much of placental transfer of materials is effected by active transport rather than passive diffusion, the physical permeability of the barrier is not of primary importance. Even in diffusion, the permeability of

Table 6.1 Classification of placental types after Grosser.

	Maternal			Fetal		
	endothelium	connective tissue	epithelium	trophoblast	connective tissue	endothelium
epitheliochorial	+	+	+	+	+	+
syndesmochorial	+	+	−	+	+	+
endotheliochorial	+	−	−	+	+	+
haemochorial	−	−	−	+	+	+
haemoendothelial	−	−	−	−	±	+

epitheliochorial	pig, horse
syndesmochorial	sheep, goat, cow
endotheliochorial	cat, dog
haemochorial	man, monkey
haemoendothelial	rabbit, guinea pig, rat, mouse

one cell layer rather than the thickness overall might be the factor limiting placental transfer.

Even in purely anatomical terms, Grosser's classification needs modification. Close examination shows that the "syndesmochorial" placentae of sheep and cows are in fact epitheliochorial, and that in apparently haemoendothelial placentae, one or more layers of chorion may surround the capillaries. There is therefore good reason to believe that Grosser's "haemoendothelial" category is not valid, and that rodents and lagomorphs have a form of haemochorial placenta. A further complication is that in the course of gestation the number of intervening tissue layers may not be constant. The rabbit placenta, for instance, is initially epitheliochorial, passes through all intermediate categories, and ends as a haemochorial (if not "haemoendothelial") type.

6.3 Metabolic relations between mother and fetus

Glucose appears to be the principal source of energy for the developing fetus. A full-term human fetus, for instance, uses nearly 30 g glucose per day. The rate of passage of glucose across the placenta is too great to be accounted for by passive diffusion. Although the concentration of glucose on the maternal side of the placenta is always greater than that in the fetus, diffusion down this concentration gradient is actively assisted.

Such a high rate of carbohydrate metabolism naturally means a high rate of oxygen consumption. Oxygen consumption by a sheep fetus may be 6–8 ml/kg/min. Oxygen enters the fetus, as CO_2 leaves it, by simple diffusion, facilitated by the countercurrent apposition of fetal and maternal placental circulations.

Of the other nutrients required to cross the placenta, amino acids are transported actively, against often considerable concentration gradients; the concentration of some amino acids in fetal blood may be more than tenfold greater than in the mother. In contrast, urea leaves the fetus by simple diffusion, to be eliminated in the maternal kidneys. In the sheep, the rate of loss of urea in this way may be 0·54 mg/min/kg fetal weight; higher than the rate of nitrogen excretion in a newborn lamb.

In view of the considerable activity of the placenta as an organ of transport, as well as an endocrine organ, it is not surprising to find that its own metabolic requirements, as distinct from those of the fetus, are considerable. It is difficult to separate placental from fetal metabolism. However, the metabolic demands of the placenta are indicated by the fact that the placental oxygen consumption is even higher than that of the rapidly-growing fetus; in the sheep it has been estimated at 25–35 ml/kg/min.

In order to maintain the fetus and placenta, the pregnant female must adapt physiologically. Nitrogen retention increases in pregnancy, water and electrolyte metabolism alter, and various nutrients are mobilized. Although, in the human at least, many of these physiological changes have been carefully monitored, the mechanisms which control them are not well understood. HPL has been implicated in these preparations for pregnancy, and certainly produces changes in protein and carbohydrate metabolism (e.g. elevating blood glucose levels) related either to the mother's ability to cope with pregnancy, or to preparations for later lactation.

CHAPTER SEVEN

BIRTH AND LACTATION

7.1 The length of gestation

The length of the gestation period of eutherian mammals is related to adult size: larger mothers tend to gestate their young for longer. Large maternal size is, in general, associated also with fewer young in each litter, and a decrease in the total weight of the litter relative to that of the mother (Table 7.1).

However reasonable these relationships may seem, their evolutionary significance is not self-evident. Probably several factors have operated independently. For instance, large animals tend not to give birth and conceal their young in burrows, hence greater maturity of the young at birth might enhance their survival prospects. Extension of the gestation period to produce young which might be larger (and therefore fewer) is one way in which greater maturity at birth might be achieved. For the most part, however, the available information is too sparse for such assessment of selective advantage to be more than speculative.

7.2 Parturition

During pregnancy, the decidualized endometrium is maintained by progesterone of either luteal or placental origin. The myometrium (the

Table 7.1 Maternal weight, gestation lengths, litter sizes and relative litter weights of eutherian mammals

Maternal weight (kg)	Gestation (days)	Litter size (mean)	Mean litter weight (kg)	Litter weight as % of mother's weight
0·1	30·9	3·4	0·0077	27·9
0·1 −	52·0	3·9	0·035	14·4
1 −	76·1	3·4	0·401	11·7
10 −	237·0	1·3	3·77	8·7
100 −	271·8	1·8	14·5	5·2
1000 +	384·8	1·0	736·6	3·3

smooth-muscle layer surrounding the endometrium) is also influenced by steroids. Oestrogen stimulates an increase in the bulk of smooth muscle and, among other metabolic effects, increases its content of ATP. In a non-pregnant animal, oestrogen treatment enhances rhythmic contractions of the uterus, as well as dilating uterine blood vessels and increasing electrical conductance and the response of the uterus to stimulation by oxytocin. The effect of oestrogen during pregnancy is therefore to prepare the contractile element of the uterus to expel the fetus at term.

In a pregnant animal, contractions of the uterus appear to be suppressed by progesterone. In the rabbit, progesterone has been shown to lower the excitability of the myometrium and its responsiveness to oxytocin. This may be achieved by elevating the resting potential of myometrial cells, and so inhibiting propagation of action potentials throughout the tissue. The importance of this "progesterone block" is shown by the fact that after removal of the ovaries (which in the rabbit are the only source of progesterone; section 6.1.6) contractions of the uterus increase and eventually, by their mechanical action, kill the fetus within.

If progesterone suppresses uterine contractions, then withdrawal of progesterone would be a suitable trigger for birth. In many species, a decline in circulating progesterone levels does precede birth (figure 7.2), although the decline is often not acute, and it is difficult to assess the significance of circulating progesterone levels in a species where placental progesterone is being applied directly to the uterus. Postponement of birth should result from artificially prolonging high progesterone levels. In the rabbit and rat, this is the case. In the cow, progesterone injections have no effect, although normal delivery is accompanied by a decline in progesterone levels, and a genetic strain of cattle is known in which persistent high levels of progesterone are associated with delayed birth. The significance of progesterone levels clearly differs from species to species. It is possible that the ratio of progesterone to oestrogen is of greater significance than the level of progesterone itself. Other factors are certainly involved in some species; for example, relaxin (section 6.1.3) has been shown to suppress myometrial activity in the guinea pig and other species.

In view of the tendency, in the course of gestation, for increasing endocrine autonomy of the fetus (section 6.1), it seems reasonable to suspect a fetal role in causing the changes in oestrogen and progesterone levels associated with birth. There is now good evidence that the timing of birth may be largely under fetal control.

In the first instance, this was suggested by investigation of naturally-occurring anomalies of reproduction. Sheep grazing on some mountain pastures in Idaho, at certain times of year, failed to give birth at the expected time. After a considerable delay, the ewes often died during the delivery of lambs two or three times as heavy as normal. The only major anatomical abnormalities in these lambs were underdeveloped adrenals and a malformed, or absent, pituitary. These defects were eventually attributed to an alkaloid produced by the plant *Veratrum californicum* on which the ewes had been grazing during early pregnancy.

Prolonged gestation occurs also with women pregnant with anencephalic fetuses, and in a strain of Guernsey cattle with a genetically-determined malformation of the pituitary.

As in all of these cases the mother was normal, it seems likely that the fetus has some control over timing of parturition, and that normal function of the fetal pituitary is essential for normal birth. An obvious test of this hypothesis is to destroy the pituitary of a fetus in the uterus artificially; this should result in the postponement of birth.

Liggins and others achieved this tricky technical feat. By electro-coagulation it proved possible to destroy the pituitary of fetal lambs. When more than about 70% of the pituitary was destroyed, pregnancy was extended for at least 37 days beyond the normal gestation period of 147 days; surgical delivery showed that the lambs were still alive at this time. Involvement of the fetal pituitary in birth was therefore established.

As with the Idaho sheep, these artificially hypophysectomized fetuses showed underdevelopment of the adrenal cortex. Control of parturition might therefore be mediated by pituitary adrenocorticotrophic hormone (ACTH) stimulating the adrenal cortex. This was confirmed by two experiments. Surgical removal of the adrenals also prolonged pregnancy, in this case until Caesarian sections were performed 14 days after the expected date of birth. Destruction of the adrenal medulla did not have this result.

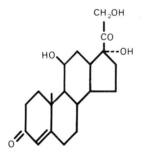

Figure 7.1 Cortisol.

Conversely, injecting ACTH into a hypophysectomized fetus reversed the effects of hypophysectomy, increasing the size of the adrenal cortex and inducing birth.

It can be concluded, therefore, that the sequence of events in normal parturition probably involves release of pituitary ACTH which stimulates the fetal adrenals.

The adrenal cortex synthesizes many steroids, any of which might constitute the next link in the chain of events leading to birth. The obvious experiment is to see which of the corticosteroids injected into an adrenalectomized, or hypophysectomized, lamb fetus have the ability to elicit birth. Only glucocorticoids were found to be effective, among the most effective being cortisol (figure 7.1).

Cortisol acts on the placenta in a variety of ways. It appears to induce two enzymes responsible for the conversion, respectively, of progesterone to 17α-hydroxyprogesterone, and of the latter to androstenedione. This will stimulate the conversion of progesterone to oestrogen (see figure 4.6) and will therefore alter the balance between progesterone and oestrogen in favour of the latter. Uterine contractions would then be allowed.

Cortisol is also thought to stimulate the production of prostaglandin $F_{2\alpha}$ (figure 2.10) probably by the endometrium. This is a powerful smooth-muscle stimulant and is used clinically in the artificial induction of labour in women. It presumably plays a similar role in normal circumstances.

In addition to its action on the myometrium, prostaglandin $F_{2\alpha}$ stimulates the release of free oestrogen, adding to the prenatal oestrogen rise. Oestrogen may act back on the placenta to increase prostaglandin levels still further. The result is a positive feedback between oestrogen and prostaglandin, and an explosive rise in both (figure 7.2).

The rise in plasma cortisol (figure 7.1) may be accelerated by a further positive feedback mechanism. This is suggested by an experiment in which parturition was induced by injecting one of a pair of twin lamb fetuses with cortisol. Cortisol levels rose also in the other twin. Maternal cortisol levels did not rise, so it is unlikely that cortisol passed from one fetus to the other across both placentas and through the maternal circulation. A more probable explanation is that raising the cortisol level in the first fetus initiated uterine contractions, and that the second fetus responded to this by releasing increased amounts of cortisol into its own circulation. Raised cortisol appears to cause uterine contractions which raise cortisol levels still further; another positive feedback loop.

The role of oxytocin in the control of birth is not altogether clear. If it has a significant role in promoting uterine contraction in normal labour, a rise

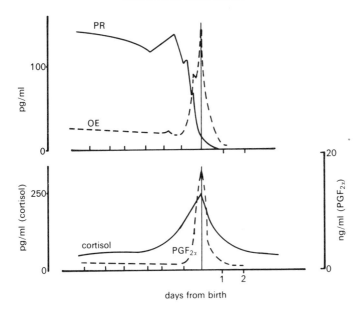

Figure 7.2 Hormone levels and birth in the sheep. *Upper diagram*; progesterone (continuous line) and oestrogen (broken line); *lower diagram* cortisol and prostaglandin $F_{2\alpha}$. All hormones measured in the uterine vein of the ewe, except cortisol which refers to fetal plasma concentration. (From Bedford *et al.*, Nathanielsz *et al.* and Thorburn *et al.* in Perry (1972)).

in maternal circulation might be expected. In the species so far studied (including sheep) oxytocin levels show little change in labour, but rise sharply during actual delivery; in the cow, from 30–150 μU ml^{-1} to 350–1000 μU ml^{-1}. As stimulation of the cervix and vagina has in many species been shown to elicit the reflex release of oxytocin, it seems likely that pressure of the fetus on the uterine cervix during labour is responsible for the increase in circulating oxytocin then. Reflex release of oxytocin in the goat is enhanced by oestrogen and inhibited by progesterone; if this is generally true, it fits neatly with other hormonal events of labour.

In women too there have been reports of high levels of oxytocin in labour. However, recent results fail to confirm a consistent large increase in output at the time of delivery, and suggest intermittent release at regular intervals, resembling the pattern found during lactation. Oxytocin release during labour may therefore be a prelude to lactation. Of possibly greater significance is oxytocin production by the *fetal* pituitary. The levels measured in the umbilical artery during labour are consistently greater than those in the umbilical vein (2·7 μU ml^{-1} cf. 1·5 μU ml^{-1}), indicating net passage from fetus to placenta. The fetal origin is confirmed by a total lack in either artery or vein in an anencephalic fetus. Application of oxytocin locally to the uterine wall may offset the relatively low concentrations involved. However, as the human placenta is rich in oxytocinase, which should prevent passage of fetal oxytocin to the myometrium, the function of this oxytocin may be within the placenta, possibly to aid in returning to the newborn the 100–150 ml blood contained within the placental blood vessels.

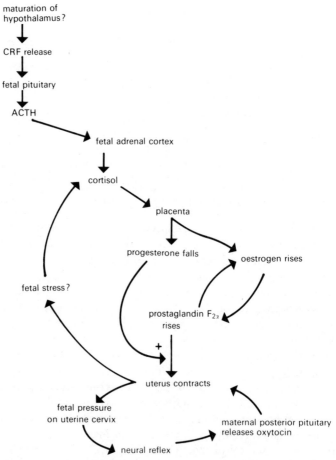

Figure 7.3 Possible interactions in the control of parturition, based mainly on evidence from sheep.

This model of the control of parturition, which is summarized in figure 7.3, has been developed as the result of research mainly on the sheep, and will certainly not apply exactly to all species. The differing significance of progesterone withdrawal has already been refered to; other interspecific differences will also emerge when more species have been studied. Species in which the luteal progesterone is essential until term (e.g. goat) will probably differ from those (e.g. sheep) where placental steroids replace luteal during pregnancy; the prostaglandin rise may act luteolytically to induce birth. In primates, the enzymes which start the conversion of progesterone to oestrogen are in the adrenals and not in the placenta. Here, the link between adrenals and placenta is not cortisol, but androgen.

Many similar variants of the postulated scheme will doubtless be discovered. One feature which is likely to be universal, however, is the existence of positive feedback mechanisms. Once labour has started, it must continue, and be completed as soon as possible. Positive feedback loops are a ubiquitous means of accelerating a process once begun, and are thus a vital part of the control of parturition.

Still something of a mystery is the stimulus which originally activates the fetal pituitary to release ACTH. A hypothalamic releasing factor (corticotrophin releasing factor, CRF) is the immediate stimulus to ACTH release. One suggestion is hypothalamic maturation during the last few days of pregnancy. The temperature of the fetal brain is 0·4–0·8°C higher than that of the maternal blood; once the thermorecptors of the hypothalamus mature they detect this mild thermal stress and respond with the release of CRF, an eventual consequence of which is parturition.

7.3 Adaption to the outside world

During pregnancy, the fetus lies in a fluid medium whose chemical composition and temperature are closely regulated. The placenta functions as a source of oxygen and nutrients, and as a means of disposing of carbon dioxide and nitrogenous waste. At birth, this undemanding situation alters abruptly. Gas exchange now takes place across the lungs, the newborn animal's own kidneys become the only means of disposing of waste nitrogen, nutrition (in the form of milk) is available only intermittently, and the ambient temperature may be 20 or more degrees below that inside the uterus. Clearly major physiological adaptations are required.

The circulation of the newborn is redirected to take account of the fact that the lungs, and not the placenta, are now the site of gas exchange. This involves closure of the aperture which in the fetal heart permits passage of blood from the right to the left atrium, and of the ductus arteriosus which takes blood from the right ventricle directly to the body. The former probably results from a fall in pressure to the right atrium causing a flap-like valve to obscure the aperture, while the ductus arteriosus closes in response to a rise in oxygen tension.

Before birth the alveoli of the lungs are collapsed. The pressure required to expand them with air is proportional to the surface tension at the

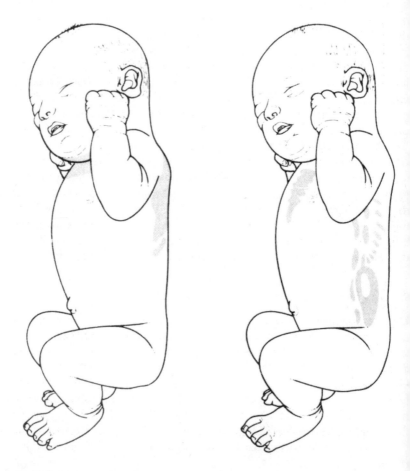

Figure 7.4 Distribution of brown adipose tissue in the newborn human. From Dawkins, M. J. & Hull. D. (1965); The production of heat by fat, *Scientific American*, **213**/2:62–7. (p. 63).

alveolar surface (and inversely proportional to the diameter of the alveolus). Lung expansion after birth is facilitated by the secretion of a surfactant (detergent) which lowers surface tension within the alveoli. This has been identified as a lipoprotein, and its secretion in the lung can be induced by injection of glucocorticoids, indicating a further effect of the prenatal rise in cortisol (section 7.2). The "respiratory distress syndrome" of the newborn human results from a deficiency of the lung surfactant.

Nutritionally, the fetus must adjust from a diet consisting largely of glucose supplied through the placenta, to milk which is relatively low in carbohydrate and rich in lipid and protein. As it is ingested and absorbed in the gut, secretion of digestive enzymes must commence shortly after birth.

Finally, the neonate must acquire control of its body temperature. The metabolic costs of thermoregulation are reduced by the proximity of the mother and of litter-mates, and frequently by some form of insulating nest. Nevertheless, much of the time energy must be expended to maintain body temperature by metabolic means. Most newborn animals are unable to shiver, so heat must be generated by other means. The principal source of heat production in the newborn mammal are deposits of highly thermogenic brown adipose tissue (BAT). BAT is strategically situated around the vital organs and the major arteries supplying them, such as the carotid artery (figure 7.4). Mobilization and oxidation of fatty acids, and the uncoupling of oxidative phosphorylation in the numerous mitochondria result in the production of much heat. Figure 7.5 shows the rise in oxygen consumption

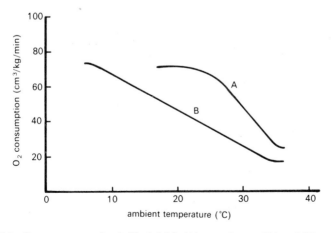

Figure 7.5 Oxygen consumption (ml/kg/min) in (A) a new born rabbit and (B) a newborn guinea pig, at various temperatures. (From Open University (1974)).

of a newborn rabbit as ambient temperature falls and its oxidation of thermogenic fat reserves increases.

7.4 Lactation

The newborn mammal, no less than the fetus, is totally dependent on its mother. Lactation is as vital to reproductive success as anything which occurs before birth. The mammary glands must therefore be fully developed, and ready to secrete, by the time of birth.

The structure of the mammary gland is similar in all mammals. Organization of the glandular tissue into numerous alveoli ensures a high surface area and consequently allows rapid secretion (figure 7.6). Surrounding the alveoli are myoepithelial cells which, by their contractions, expel milk into fine milk ducts. These in turn unite to form

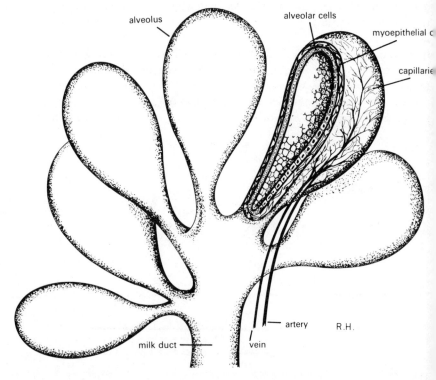

Figure 7.6 Diagram showing the structure of the alveoli in the mammary gland of a goat.

progressively larger ducts; eventually major milk ducts carry the accumulated milk through a nipple to the exterior. The way in which the ducts connect differs from species to species, as do the number and disposition of the mammary glands themselves.

Mammary gland tissue grows throughout life, in proportion to growth of the whole body. The major phase of growth follows conception, when the alveoli become fully developed and the ducts highly branched. Removal of relevant endocrine organs and replacement therapy have established the hormonal control of this growth in the rat (which, as ever, may not be typical). Oestrogen, pituitary growth hormone, and corticosteroid act synergistically to promote duct growth, while for full development of the alveoli progesterone and prolactin are necessary in addition. Hormones involved in the maintenance of pregnancy are thus equally important in preparing for lactation.

After birth, two secretions are successively formed in the mammary glands, colostrum and true milk. The former has a lower content of lipid than milk, and a higher protein content (Table 7.2). In many species it is a major source of the antibodies passively acquired by the fetus from its mother (section 9.6).

Table 7.2 The major constituents of human colostrum and mature milk. Comparable figures for the milk of a variety of other species are also shown. All figures are in g/100 ml.

	human colostrum	human milk	cow milk	rabbit milk	elephant milk	blue whale milk
water	89	88	87	67	78	43
fat	2·9	3·8	3·7	18·3	11·6	42·3
protein	2·7	1·0	3·4	13·9	4·9	10·9
lactose	5·3	7·0	4·8	2·1	4·7	1·3
ash (mineral salts)	0·3	0·2	0·7	1·8	0·7	1·4

A high proportion of milk consists of fat, present as a suspension of droplets of about 3 μm diameter, comprising a complex mixture of phospholipids, fat-soluble vitamins, carotenoids and other substances. The carbohydrate content is largely the disaccharide lactose. Protein constituents fall into two categories, caseins and whey proteins. Caseins are phosphoproteins of molecular weight 20,000–27,000. Digestion of casein is facilitated by precipitation in the gut of the young animal, either enzymatically or by the low pH in the stomach. Calves, for instance, secrete the enzyme rennin which curdles milk in the stomach. As casein molecules in milk are associated with calcium and phosphate ions, they are probably

H

important in supplying to the neonate materials necessary for bone growth, as well as amino acids for protein synthesis.

Many of the whey proteins appear simply to be useful sources of amino acids, but some have specific functions as well. Lactoferrin, present in the milk of a number of species, binds strongly to iron and may be important in transferring it to the young animal. Sequestration of iron may also be the basis of lactoferrin's bacteriostatic action; it is an important factor in reducing the susceptibility of the newborn to gut infections.

Another of the whey proteins, α-lactalbumin (actually a globulin), also has an important role. Lactose is synthesized in the mammary gland by an enzyme, "lactose synthetase". This consists of two dissociable subunits, A and B. A is a widely distributed enzyme, normally catalysing a different reaction. In the presence of protein B, its specificity is altered; as a result, lactose is formed (figure 7.7). Protein B has been shown to be identical with α-lactalbumin.

As lactose is the major osmotic component of milk, lactose secretion is probably necessary for water movement across the secretory epithelium. Without this solvent movement, the other solutes of milk could not be transported into the alveoli. Consequently, if lactose is not secreted, milk production does not take place. Milk secretion is therefore dependent on the existence of functional "lactose synthetase". Synthesis of both subunits of the enzyme is probably induced by prolactin, in the presence of insulin and cortisol. Formation of α-lactalbumin, however, is selectively inhibited by progesterone. Hence although the A subunit is present in increasing

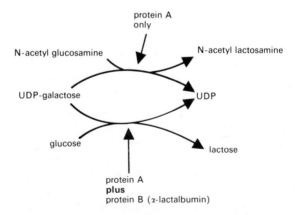

Figure 7.7   The control of lactose synthesis during the production of milk. (From Mepham (1976))

amounts in mammary gland tissues through pregnancy, α-lactalbumin remains at low levels until the decline in progesterone which heralds birth; α-lactalbumin concentration is therefore rate-limiting in lactose synthesis and consequently in milk secretion. The decline in progesterone levels at the end of pregnancy initiates milk secretion as well as labour.

Physical stimulation of receptors in the nipple ends leads to the reflex release of oxytocin from the posterior pituitary of the mother (figure 2.13). Oxytocin, as a stimulant of smooth-muscle contractions, causes the myoepithelial cells to squeeze the milk out of the alveoli.

In many species, no milk can be obtained by the young animal without the suckling reflex. Oxytocin release can also follow a conditioned stimulus; for example, in a cow oxytocin release may be evoked by the entry of a milker, and in a woman by hearing the cry of her baby. Suckling also causes a release of prolactin by the anterior pituitary, which makes it possible for the rate of milk synthesis to be regulated by demand.

Lactation imposes a considerable metabolic burden on the mother. A woman producing 400 ml of milk a day (not an excessive amount; the record is 5·95 l/day!) is supplying 15 g fat, 4 g protein, and 28 g lactose per day (compare with the 30 g glucose/day used by a full-term human fetus). In dairy cattle, the mammary glands (about 5% of the total body weight) account for two-thirds of all glucose available for the body. It is not surprising that female mammals cannot sustain lactation simultaneously with pregnancy, and that mechanisms have evolved to avoid this happening. Lactational delay of implantation in mice (section 5.4) defers the need to support a brood conceived at a postpartum oestrus. In the rat, suckling more than 8 young may also cause death and resorption of implanted embryos.

In other species, including our own, lactation supresses ovulation (sections 10.2 and 10.6). The mechanism which postpones the return to normal fertile cycles is not yet clear. High prolactin levels, typical of lactation, correlate with the period of infertility in rat and monkey. Injected prolactin also suppresses ovulation, but only during suckling. It appears to act by reducing the sensitivity of the pituitary to LRF, so preventing the LH rise necessary for ovulation. There is also evidence of an altered response of the ovary to FSH.

CHAPTER EIGHT

SEXUAL DEVELOPMENT AND DIFFERENTIATION

8.1 Chromosomal sex

In most species of mammal, the sexes are readily distinguished on the basis of their chromosome complement. In the female, all chromosomes are paired, the sex-chromosomes (X) as well as the autosomes; the male has a single X and a single Y chromosome (figure 8.1). Birds and some of the lower vertebrates show the reverse, with the female as the heterogametic sex

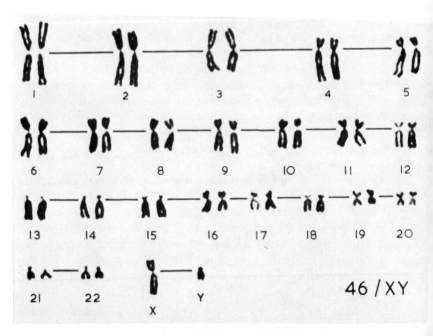

Figure 8.1   The human male karyotype: in women two X chromosomes are present and the Y chromosome absent. (Original photograph by L. Butler).

and the male with a pair of identical sex chromosomes. Among the mammals, translocation and other cytogenetic accidents have modified the pattern in a few species. In several species of mongoose (*Herpestes*) the Y chromosome is attached to an autosome, and in the common shrew (*Sorex araneus*) two Y chromosomes are found (Y_1 and Y_2). Nevertheless, with these minor exceptions, the male is universally the heterogametic sex among mammals.

There are two possible views of the significance of the chromosomal differences between the sexes. Femaleness could result from the possession of two X chromosomes rather than one; or maleness from the presence of a Y chromosome and not from the lack of a second X. Non-disjunction of chromosomes can result in abnormal sex chromosome complements which help to elucidate this point; Table 8.1 lists some recorded human sex chromosome complements. The number of X chromosomes is irrelevant to sexual phenotype. The presence of a Y chromosome is sufficient to establish a male phenotype, and in its absence the phenotype is female.

The Y chromosome is typically much smaller than the X. It can therefore contain less DNA and therefore less genetic information. As a means of compensation for a greater "dose" of X-linked genes, female mammals inactivate all but one of the X chromosomes in every cell. The inactivated chromosome is cytologically detectable, as a "Barr body", cells of a normal female having a single Barr body, those of a XXX female two, and so on. Inactivation occurs early in embryonic development. In each cell, chance decides whether the paternal or maternal X chromosome is inactivated. Whichever it is, all descendants of a cell respect the initial decision;

Table 8.1 Some recorded human sex chromosome complements, and the sexual phenotypes which result from them.

Male	*Female*
XY (normal male)	XX (normal female)
XXY (Klinefelter's syndrome)	X (Turner's syndrome)
XYY	XXX
XXYY	XXXX
XXXY	XXXXX
XXXXY	

XXY men are generally normal in appearance, but sterile; a proportion show breast development and underdeveloped secondary sexual characteristics. Women with a single X are sterile, but those with XXX are fertile, and their offspring normal. In general, the more sex chromosomes, the more extensive the physical deformities and the greater the mental impairment.

remarkable when it is considered that the unexpressed chromosome is not totally inert, but must be replicated at each mitotic division of the cell.

Many genes are sex-linked, that is carried on the X chromosome. Y chromosomes appear virtually devoid of conventional genetic information. Recently it has been found that some 50% of the Y chromosome DNA consists of simple repetitive sequences. One possible explanation is that the significant section of the DNA—a small region near the centromere—is so small that some "padding" is needed to make it viable as a chromosome. The low information content of the Y chromosome, and the developmental flexibility of the reproductive system (see the later sections of this chapter) make it clear that the Y chromosome cannot specify in detail the structures required for morphogenesis of a normal male. Its function must be as a controlling element, deciding which of two alternative patterns of sexual development should be initiated.

8.2 Gonadal sex

If a Y chromosome is borne, the gonads are testes; if not, then ovaries are found. Other differences in sexual phenotype are contingent on the nature of the gonads. This clear genetic determination of gonadal sex in mammals differs from the more labile sex determination of some other vertebrate groups, where sex reversal of the gonads is relatively easily obtained by hormone treatment. It has been suggested that the more definite genetic control in mammals is an adaptation to viviparity, where a male fetus might otherwise run the risk of accidental sex reversal by exposure to hormones from his (female) mother. This suggestion may be valid, although hormones do not freely pass into the fetus during pregnancy, and viviparity is not the exclusive prerogative of mammals.

In order to understand how the presence of a Y chromosome might determine gonadal sex (and subsequently the sex of the genital ducts, behaviour, and so on) it is necessary to consider the embryological origin of the gonads and germ cells.

Gonadal primordia first appear as small ridges on the surface of each pronephros, consisting of a small area of mesenchymal cells surrounded by coelomic epithelium. The primordial germ cells originate elsewhere, probably in the epithelium of the yolk sac, near the allantois. They then migrate, by amoeboid motion, through the hindgut endoderm to the gut mesentery, and eventually reach the gonadal primordia. Chemotaxis may help to guide them to their target. During the migration, multiplication occurs; in the mouse when the primordial germ cells are first recognizable (early on the 8th day of pregnancy) they number 100–150; this number has trebled by the following day, and by the 12th day about 5000 enter the gonad.

Until entry of the germ cells into the gonadal primordium, the sex of both can be told only by chromosome analysis. Subsequently, the first signs of sexual differentiation appear. If the fetus is genetically female, the bulk of the germ cells remain within the epithelial cortex of the gonad. The medulla will eventually diminish and the future ovary progressively differentiates from the cortex, with cells of epithelial origin induced to surround the germ cells (gonocytes) to form ovarian follicles. If the fetus is male, primordial germ cells penetrate the medulla, the mesenchyme develops into Leydig and Sertoli cells and becomes organized into "sex cords" from which seminiferous tubules eventually result.

Several theories have been advanced to explain how the Y chromosome could control the emergence of medulla rather than cortex as the main constituent of the future gonad. Cells containing XY chromosomes contain slightly less DNA than those with XX; perhaps as a consequence they divide faster and more cells are formed in a given time. At the stage when little histological difference is apparent between male and female gonad rudiments, the future testes are in fact larger than future ovaries. A slight quantitative difference in growth rates might seem inadequate to explain the major quantitative differences which arise later. However, the low temperature experienced by a genetically female gonadal primordium after transfer to a scrotum does suppress the cortical region and induce the medulla to develop as a testis. Possibly the low temperature affects relative growth rates of cortex and medulla.

Alternatively, the Y chromosome might produce some inducer substance determining maleness. It is necessary therefore to identify the primary gene product of the male-determining region of the Y chromosome. One possible candidate is a cell surface component known as the H-Y antigen. This occurs on the cells of male mammals and can elicit rejection of skin grafts from male to female in circumstances where the reciprocal graft would be accepted. Evidence implicating H-Y in sex determination is accumulating, but is largely circumstantial. H-Y is found only when a Y chromosome is present. A mutation of mice is known which results in deficency of cytoplasmic and nuclear androgen receptors. The mutant gene (known as Testicular feminization, Tfm) is sex-linked, so that a Tfm/Y animal has rudimentary testes, but is otherwise phenotypically female. Despite feminization, H-Y is still present. It is therefore dependent on a Y chromosome and not on a male phenotype.

If the H-Y gene is not identical with the male-determining region of the Y chromosome, it is extremely closely linked to it. This is shown by another mutant, "sex-reversed" (Sxr), which probably takes the form of a translocation of a part of the Y chromosome, too small to be detected cytogenetically, onto one of the autosomes. A mouse with two X chromosomes, but carrying Sxr, is male in phenotype and carries H-Y antigen.

Finally, H-Y is evolutionarily highly conservative, always being found in the heterogametic sex; that is, in association with the sex-determining chromosome. This implies that it has a specific function, and that its association with sex determination is unlikely to be fortuitous.

Regardless of phenotype, then, H-Y antigen is associated with development of a testis. One clue as to how it might possibly act comes from the study of freemartins. In some species, particularly cattle, the placentae of twins of opposite sex may grow together to such an extent that a common placental circulation develops and exchange of cells takes place. In such cases, normal sexual development of the female twin is impaired. Although genetically female, she may develop testis-like gonads and male accessory structures.

Although freemartinism is a complex and variable phenomenon, an important feature is now thought to be infiltration of the primordial gonad of the female twin by genetically male cells. In the putative ovary, a relatively small number of XY cells may exert a disproportionate effect; fewer than 10% male cells are found in extensively virilized ovaries. It has been proposed that H-Y antigen is released from the surface of XY cells, coats the surface of the majority of genetically XX cells of the medulla, and by so doing induces them to form a testis. In the presence of H-Y antigen, XX cortical cells degenerate.

This proposal for the action of H-Y is speculative, and the evidence for it largely circumstantial. Other substances may be implicated in sexual development, and H-Y may prove to correlate with male sexual development but not determine it. It seems clear, however, that some gene product of a Y-linked gene must participate, and must have many of the properties possessed by H-Y.

Survival of germ cells depends on the genotype of the primordial gonad they enter. A dominant autosomal gene in the goat causes, in the female, development of the medulla instead of the cortex. XX germ cells thus enter a "testis", also of sex karyotype XX. Normal germ cell development proceeds until day 120–130 of the gestation period of 150 days; by birth all XX germ cells have degenerated. Conversely, in a mouse chimaera consisting of a mosaic of cells of the karyotypes 39,X and 41,XYY all spermatogonia and spermatocytes are found to be 41,XYY. Germ cells with the 39,X constitution are perfectly viable, since in a homogeneous mouse of the same constitution they enter the gonadal cortex and eventually form oocytes. In the chimaera, the presence of cells carrying Y chromosomes converts the mosaic gonad into a testis in which only the 41,XYY germ cells appear able to survive.

An effect of germ cells, as opposed to somatic cells, on the course of gonadal development is indicated by the development of ovaries in human fetuses with a single X chromosome (Turner's syndrome). As long as germ cells are present, the presumptive ovaries continue to develop. At a certain stage, the germ cells degenerate, and regression of the ovaries soon follows.

The interactions between germ cells and primordial gonads are therefore subtle and complex, but poorly understood.

8.3 Differentiation of the genital tract

Following establishment of the sex of the gonads, differentiation of the genital tract begins. Like the gonads, the mammalian reproductive tract is initially bipotential. Parallel Wolffian and Müllerian duct systems develop. During normal development of a genetically male fetus, the Müllerian ducts degenerate and virtually disappear; the Wolffian system is stabilized and differentiates into epididymides, vasa deferentia, seminal vesicles and other features of the mature male. Conversely, in the female the Wolffian ducts are suppressed and the Müllerian ducts develop into oviducts, uterus and associated structures (figure 8.2, Table 8.2).

Which duct system persists and which degenerates might be decided by the sex of the gonad, since gonadal differentiation precedes duct differentiation. As emergence of a testis from a primordial gonad takes place earlier in development than appearance of a recognizable ovary, it is

Table 8.2 Fate of the Wolffian and Müllerian duct systems in normal and experimentally treated male and female treated rabbit fetuses.

	Müllerian	Wolffian
normal male	−	+
normal female	+	−
castrated male fetus	+	−
ovariectomized female fetus	+	−
male castrated + androgen	+	+
male + cyproterone acetate	−	−

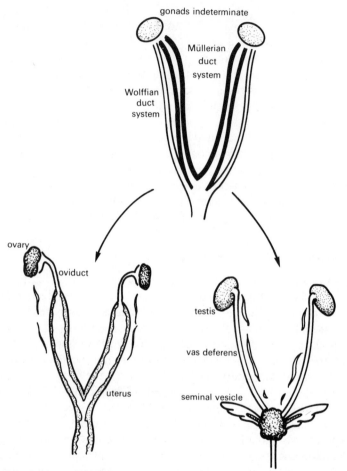

Figure 8.2   Origin of the male and female genital tracts from the sexually indeterminate stage. (From Jost (1960)).

possible that the presence or absence of a testis is the crucial factor, rather than whether an ovary is present or not. This would parallel the situation in gonad differentiation, where the Y chromosome intervenes to impose maleness on a gonad which would otherwise become female.

If a male rabbit is castrated within the uterus before the ducts have differentiated, the Wolffian ducts do degenerate and the Müllerian ducts are preserved to form a recognizable uterus. Ovariectomy of a female fetus similarly results in the formation of a uterus. This indicates that the presence of the ovary is irrelevant to the final sex of the genital tract. It is the presence of the testis which is decisive; the testis imposes maleness on an initially bipotential duct system (Table 8.2).

Replacement of one ovary of a rabbit fetus by a testis results in the unilateral suppression of the Müllerian duct system and maintenance of the Wolffian duct, confirming the action of the testis.

It seems reasonable to suggest that early androgen secretion of the fetal testis exerts the masculinizing influence. To test this, male rabbit fetuses were castrated in the uterus and later injected with androgens, to substitute for the missing testes. This treatment stabilized the Wolffian duct system and allowed it to differentiate. It did *not* suppress the Müllerian ducts; the result was the persistence of *both* duct systems in parallel. Cyproterone acetate, a substance known to be a specific antagonist of the action of androgens, had the opposite effect; maintenance of the Wolffian system was blocked, but suppression of the Müllerian system allowed, and the treated animals had testes but no genital duct system of either sex (Table 8.2).

Masculinization of the rabbit genital tract by the testis must therefore be effected by two substances. Androgens stabilize the Wolffian ducts and permit their further differentiation; a separate substance is responsible for suppressing the Müllerian ducts. This latter substance cannot be an androgen. It has not yet been identified with certainty, but emanates from the seminiferous tubules, probably from the Sertoli cells, and appears to be a protein.

In the male rat fetus, cyproterone acetate treatment suppresses only certain structures of the male duct system; not, for instance, the seminal vesicles which differentiate relatively early. Some other masculinizing stimulus must be responsible for seminal vesicle development. Probably in other species, other variants of the mechanisms will be found. It is probably generally true, however, that in the presence of a testis the genital tract will become male in nature and in its absence female, irrespective of the chromosomal sex. The presence or absence of ovaries is irrelevant.

8.4 Sexual differentiation of the brain

At birth, the reproductive system is anatomically complete. Its mode of operation, however, is not yet determined. If a male rat is castrated at birth and later, as an adult, receives an ovary transplant, then it will undergo cycles of ovulation and luteinization like those of a normal female (figure 8.3). This can readily be monitored by grafting, in addition, some vaginal

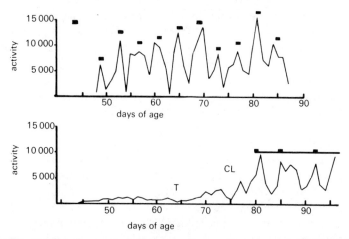

Figure 8.3 Evidence for a cyclic hypothalamus in a male rat castrated at birth. The upper diagram shows the cycles of activity (daily revolutions of an activity wheel) of a normal female rat. The black bars indicate the days of oestrus, as judged by the examination of vaginal smears. The lower diagram indicates the development of similar cycles in a neonatally castrated male rat which was given an ovarian and vaginal transplant at 63 days of age. T indicates the time of the transplant, CL the age at which corpora lutea were first detected. (From Harris (1964)).

tissue; the artificial vagina undergoes a more or less rhythmic cycle, reflecting the ovulatory cycles of the grafted ovary. Female sexual behaviour is also shown; for example, the lordotic response discussed in section 4.5. An ovary transplanted to a male rat castrated as an adult does not behave cyclically.

If similar experiments are performed on a female rat (ovariectomy at birth, and implantation of an ovary when adult) the result is the same—cycles of ovulation and luteinization, with appropriate sexual behaviour. Conversely, transplantation of a testis into a newborn female rat renders her permanently sterile. Her ovaries fail to ovulate and form corpora lutea.

These findings indicate that development of a male reproductive pattern is determined by the presence, or otherwise, of a testis. In its absence, a cyclic female pattern will result. The presence or absence of an ovary is immaterial. It is also evident that reproductive function is determined shortly after birth and that, if a testis is present during this critical period, its later removal does not permit a male rat to revert to a female reproductive pattern.

The effects of castration on reproductive function can be duplicated by injecting the antiandrogen cyproterone acetate during the critical period just after birth. Androgen, probably testosterone, is therefore responsible for the masculinizing action of the testis. Treatment of newborn female rats with testosterone (testosterone propionate is generally used) simulates the presence of a testis. When mature, these "androgenized" females fail to ovulate and form corpora lutea, and are permanently sterile. Figure 8.4 shows the time after birth for which a single androgen injection (10 μg testosterone propionate) is effective in inducing sterility in female rats. Extension of the critical sensitive period prenatally, to at least 3–4 days before birth, can be shown by injection of TP into a female fetus, or into a pregnant female (although abortion or intrauterine death is apt to result in these experiments).

Paradoxically, oestrogen has similar effects to androgen when injected into newborn female rats. This probably reflects conversion of testosterone to oestrogen in target organs (section 2.7). The ovaries of a rat are normally quiescent until some time after birth, so it is unlikely that reproductive function could be disturbed by perinatal exposure to endogenous oestrogen. Prenatal action of maternal oestrogen may, in the rat, be prevented by the presence of α-fetoprotein (AFP) which specifically binds oestrogen with a high affinity. AFP in other species (such as humans) has no such steroid-binding properties and other functions have been attributed to it (section 9.3).

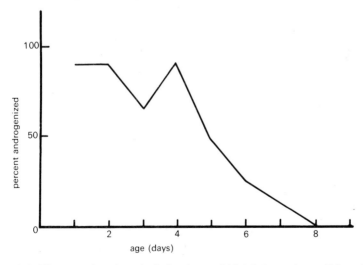

Figure 8.4   The proportion of genetically female rats which failed to ovulate at 45 days of age, following a single injection of androgen (10 μg testosterone propionate) at various intervals after birth. (From Gorski 1968).

Cyclicity of the female reproductive system is largely a property of the hypothalamus (section 4.4). Evidently, then, perinatal exposure of the hypothalamus to exogenous or endogenous androgen converts it from a female to a male mode of operation. This suggestion is supported by the observation that androgenization is prevented by the injection of a central nervous system depressant (such as a barbiturate) simultaneously with the testosterone. Given the interpretation of the function of the hypothalamus discussed in section 4.4, a simple explanation of neonatal androgenization would be the elimination of the positive feedback relationship of oestrogen with the preoptic area. This appears to determine the phasic release of LH from the pituitary; its effective elimination would leave only the tonic control of LH release, corresponding to the situation after surgical isolation of pituitary and hypophysiotrophic area from the rest of the brain (section 4.4). Adult neonatally androgenized rats do in fact fail to respond to an oestrogen injection with an ovulatory peak of LH.

Androgen treatment affects other parts of the reproductive system besides the preoptic area of the hypothalamus. In section 4.4, the modulation of pituitary responsiveness by steroids was discussed. Perinatal testosterone treatment, too, affects pituitary sensitivity in later life. In one experiment on 20-day-old rats, an injection of 250 ng LRF raised circulating LH levels from around $10 \, ng \, ml^{-1}$, to about $110 \, ng \, ml^{-1}$ in normal males, and $368 \, ng \, ml^{-1}$ in normal females. In rats which had received a single injection of testosterone propionate at 5 days of age, the comparable figures were $90 \, ng \, ml^{-1}$ and $210 \, ng \, ml^{-1}$. Androgen treatment had lowered LH response by more than 40%. Oestrogen treatment at 5 days of age had no significant effect on the response of male rats at 20 days; again, the female response was reduced, by about the same proportion as previously. Similarly, androgenization renders the ovarian follicles more refractory to LH. A higher level than normal is necessary to induce ovulation. However, the pituitary and ovarian effects of androgenization are probably of minor significance compared with the androgenization of the hypothalamus.

An androgenized female is clearly not functioning as a female. It does not follow that her reproductive function is identical with that of a male. Circulating FSH levels of androgenized females, for instance, are around $200 \, ng \, ml^{-1}$; in a normal adult female, FSH levels are $200 \, ng \, ml^{-1}$ or below, but in a normal adult male are generally greater than $600 \, ng \, ml^{-1}$. It cannot be assumed that an androgenized female is equivalent in all respects to a male, or that perinatal determination of the pattern of reproduction is so simply effected in normal circumstances that its duplication can be easily achieved artificially.

Genetically male (XY) rats are known which have an inherited deficiency in target organ receptors for androgen (sections 2.7 and 8.2). A female phenotype is the result. The effects of neonatal castration and antiandrogen treatment suggest that these rats, too, should have cyclic patterns of gonadotrophin release. They do not. A failure of neonatal androgenization in a genetic male does not necessarily permit the expression of a female pattern of gonadotrophin release. Obviously the perinatal actions of steroids on the CNS are subtle.

Other species differ from the rat in the sexual differentiation of gonadotrophin secretion patterns. In the guinea pig, which is born at a very mature stage compared with the rat, the critical period for masculinization of the hypothalamus is entirely prenatal. Primates differ more markedly.

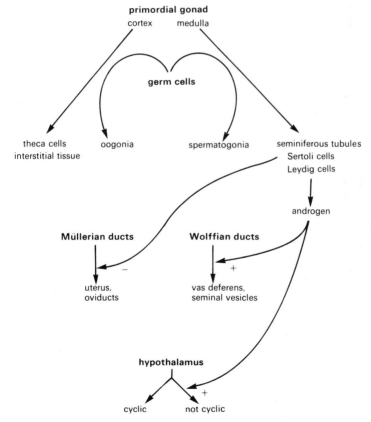

Figure 8.5 Determination of the major aspects of sexual differentation.

Female monkeys given androgen before birth have masculinized external genitalia, but may later ovulate normally. Similarly, girls with genitalia masculinized as a result of androgen production by defective adrenals may still be capable of ovulating after puberty. In monkeys, even a male castrated as an adult will respond to oestrogen treatment with a sharp rise in LH like the ovulatory surge of a female. Androgen does not suppress the positive feedback relationship in primates as it does in rodents.

Sexual differentiation therefore takes place in a series of stages. First the sex of the gonads is determined, then that of the genital tract, and finally the pattern of gonadotrophin release and sexual behaviour. Initially, the presence or absence of a Y chromosome is decisive, later the presence or absence of a male gonad. X chromosomes, and ovaries, are not relevant.

The possible course of events in sexual differentiation is summarized in figure 8.5.

8.5 Puberty

Between birth and puberty, the reproductive system undergoes maturation and becomes fully functional. The changes which culminate in sexual maturity, and their timing, are complex and not well understood.

Figures 8.6 and 8.7 show the changes in serum concentrations of pituitary gonadotrophins and gonadal steroids in humans between birth and puberty. With minor exceptions, all hormones shown (and prolactin, not shown) tend to increase throughout this period. Similar tendencies occur in other species, although with some anomalies; for example, in male rats prepubertal FSH levels are about twice those of adults, and testosterone shows a peak at about 19 days, followed by a fall and subsequent rise.

The hypothalamus plays an important part in regulating the reproductive hormones in sexually mature individuals. Most theories of the control of the onset of puberty also assign to it a significant role. According to a suggestion first advanced by Hohlweg some 40 years ago, the neonatal hypothalamus has a low threshold of sensitivity to the inhibitory effects of steroids on gonadotrophin release. From the time of birth, this sensitivity to inhibition declines; existing levels of steroids in the circulation become inadequate to prevent a rise in LRF (hence LH). Consequently, steroid levels rise until LRF production is again stabilized. Further decreases in hypothalamic sensitivity to inhibition will be followed by further rises in LH and gonadal steroids, and so on.

Castration or ovariectomy of adult mammals results in an elevation in

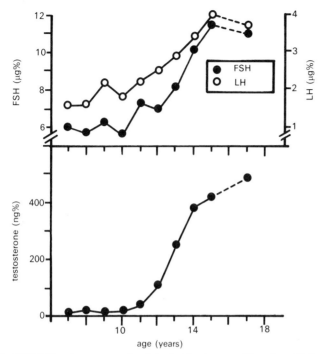

Figure 8.6   FSH, LH and testosterone levels in the blood serum of boys from birth to sexual maturity. (From Grumbach, Grave and Mayer (1974)).

circulating LH levels (sections 2.7 and 4.4), which can be reduced by steroid injection. In a number of species, the same response to castration and steroid replacement has been demonstrated between birth and puberty. From this (and other evidence) it seems clear that the feedback relationship between pituitary and gonadal hormones is established well before puberty. The concentration of steroid required to suppress the post-castration rise in LH is a measure of the sensitivity of the hypothalamus to inhibition. In some cases it has been shown that steroids are inhibitory at lower levels in prepubertal than in mature animals. This supports the hypothesis of a progressive decline in sensitivity. It has been estimated that in humans, the prepubertal hypothalamus is between six and fifteen times more sensitive than that of adults.

LH output of the pituitary in response to the intravenous injection of a given dose of LRF is greater in adult males than in prepubertal children. As the sensitivity of the hypothalamus to steroid inhibition declines, the

I

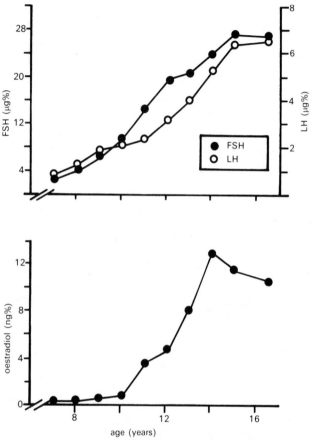

Figure 8.7   FSH, LH and oestradiol levels in the blood serum of girls from birth to sexual maturity. (From Grumbach, Grave and Mayer (1974)).

sensitivity of the pituitary to stimulation by hypothalamic LRF increases. Similarly, the response of the gonads to LH increases as puberty approaches. Whatever causes these alterations, they will enhance the effects of the changes within the hypothalamus.

The Hohlweg model therefore provides a reasonable framework for the observed rises in reproductive hormones, in terms of the adjustment of the setting of a hypothalamic "gonadostat". It does not explain the ontogeny of the *positive* feedback relationships in female mammals: the ability to respond to oestrogen with a sharp "ovulatory" rise of LH. This ability is absent in girls before puberty, which could be due either to immaturity of the neural mechanisms responsible (section 4.4) or to inadequate accumulations of LH in the pituitary. It is not known whether maturation of the positive feedback mechanism is critical in determining the time of puberty in girls, or whether the neural mechanism is mature well before puberty and awaits the rise in oestrogen to levels sufficient to evoke an LH surge, or some other maturational event.

In men, LH is released episodically and not continuously, with peaks at intervals of about an hour (section 2.7), hence their description as "circhoral". Before puberty, LH release appears to be continuous. A curious feature of the initiation of the episodic release pattern is its association with sleep. In pubertal boys (but not in boys before puberty, or in adult men or women) a rise in LH is synchronized with the onset of a particular phase of sleep. The significance of this, and indeed of the episodic release pattern itself, is not known; presumably it reflects some process of neural maturation.

Changes in the hypothalamus, although of great importance, are by no means the only factor in sexual maturation. Some control of maturation is exerted by other parts of the brain. Precocious puberty can be induced by electrolytic lesions in the amygdala, at least in female rats, and involvement of the pineal has also been established.

In humans, attainment of a certain minimum body weight may also be a precondition of puberty. Puberty in girls is more closely associated with body weight than with age, the critical body weight being 47 ± 5 kg. Loss of weight (as in anorexia nervosa) is associated with the cessation of menstruation, so the same relationship appears in reverse. It may be significant—or coincidental—that in a girl of average build a weight of 47 kg implies fat reserves of 11 kg, equivalent to approximately 99,000 calories. Pregnancy has been estimated to "cost" between 27,000 and 80,000 calories (the latter figure applying to poorly-nourished women). Menstruation will not, therefore, start or continue until the accumulated fat reserves are adequate to support pregnancy plus, say, one month's lactation (at 1000 calories per day). There will thus be little chance of pregnancy without sufficient metabolic reserves to carry it through to completion.

Attributing a great deal of significance to body weight as a decisive factor in the timing of puberty is speculative; the association may be fortuitous. Even more speculative are the postulated mechanisms whereby the brain might monitor body weight or fat reserves. It should be clear, however, that whether or not body weight is an important determining factor, the control of the pace of sexual maturation is complex and poorly understood.

CHAPTER NINE

IMMUNOLOGY AND REPRODUCTION

9.1 The immune response

Vertebrates have evolved highly sophisticated methods for the destruction of foreign material such as pathogenic bacteria. In a matter of days following exposure, the invading bacteria may be completely eliminated, and a persistent state of resistance to further infection by the same organism may develop. The infected animal is now immune to the bacteria in question.

Sperm and seminal fluid are in an analogous situation to invading bacteria, since they are derived from the male and are therefore foreign to the female. The fetus in a pregnant female is in a similar situation, and for a longer period of time; half of its genotype is paternal in origin and it is, in consequence, genetically different from the mother. Neither the transient but repeated exposure to semen, nor the prolonged and intimate contact with the fetus, appear to lead to rejection by the mother and the induction of a state of immunity—fortunately since, were this not so, reproduction would be impossible.

In order to discuss how the immune defences of the female are evaded in the course of reproduction, it is necessary first to consider how the immune system operates in other situations.

Any substance capable of provoking an immune response is described as an *antigen*. The principal requirement for a substance to be antigenic is a high degree of molecular complexity; hence, with few exceptions, proteins of molecular weight greater than 5000 are highly antigenic, as are many polysaccharides of molecular weights over 100,000. Bacteria and other cells are antigenic by virtue of protein or polysaccharide surface components. Other substances, such as lipids and nucleic acids, are only weakly antigenic.

If the antigen in solution is injected intravenously into a suitable experimental animal, it will probably be eliminated from the circulation in from 10 to 14 days. If, thereafter, the same antigen is injected again, it will disappear more rapidly, in 4 or 5 days. The recipient has developed a state of immunity to the antigen in question; it will react more rapidly and effectively to a second exposure. This is one key aspect of the phenomenon of acquired, or adaptive, immunity. By analogy with more conventional forms of memory it is often referred to as "immunological memory" and the accelerated secondary response as an "anamnestic response".

The second crucial aspect of the immune response is its remarkable specificity. Injection of a second antigen into an already immunized animal results in the primary response of elimination in 10–14 days. The immune system can therefore discriminate between a novel antigen and one of which it has previous experience. So great is the power of discrimination that in some circumstances two protein antigens can be told apart which differ only by a single amino acid. This is the basis for the discrimination between "self" and "non-self": the recognition as foreign of the antigens borne on the surface of cells transplanted from another individual, despite their close similarity to the corresponding components on the cells of the recipient.

In the case of the intravenous injection of antigen, the recipient responds with the production of antibody molecules specific to the antigen. These are proteins, collectively referred to as the immunoglobulins, or γ-globulins. The structure of an antibody molecule is shown in figure 9.1; it consists of four polypeptide chains held together by disulphide bonds in a symmetrical Y configuration. On each arm of the Y is an antigen-combining site comprising regions of both "heavy" and "light" chains whose amino-acid sequence varies according to the specificity of the antibody. The remainder of the molecule is not variable in this way.

Virtually all of the work of elucidating antibody structure has been done

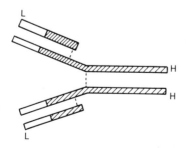

Figure 9.1 Schematic diagram of the structure of an antibody (Immunoglobulin G) molecule. H = heavy chain, L = light chain.

on Immunoglobulin G (IgG) which is the major component of induced antibody circulating in the blood. Four other classes of antibody can be distingushed on grounds of molecular size and other physical properties; their characteristics are summarized in Table 9.1. IgM is a polymer of 5 IgG-like molecules linked together with carbohydrate. It is a major component of the primary immune response, but is subsequently replaced by IgG; only IgG is detectable in a secondary response. Many antibodies which cause agglutination, such as those of the ABO blood-group system, are IgM. IgA is an inducible antibody found in small amounts in the blood, where it resembles IgG. It is also found in mucous secretions; here it is in the form of a dimer, with two IgG-like molecules joined by a "secretory piece" or "T-component". The two remaining immunoglobulin classes, IgD and IgE also resemble IgG in general structure. IgE, or reaginic antibody, mediates allergic responses. The function of IgD is unknown.

Table 9.1 Properties of the major immunoglobulin classes in humans

		Molecular weight	Sedimentation coefficient
IgG		160,000	7S
IgM		900,000	19S
IgA	serum	170,000	7S
	secreted	390,000	11S
IgD		184,000	7S
IgE		188,000	8S

When a soluble or suspended antigen enters the body, it is drained off in the tissue fluid to a regional lymph node or, if injected into the blood, to the spleen. A small proportion of the lymphocytes with which the lymph node is stocked (figure 9.2a) then start to divide; in all they undergo about eight divisions. Most of the daughter cells transform into plasma cells (figure 9.2b) with a greatly increased content of granular endoplasmic reticulum, and begin actively to synthesize protein. Culture of isolated plasma cells *in vitro* has made it possible to confirm that this protein is antibody; initially IgM, subsequently IgG. Both antibody and sensitized lymphocytes may pass out of the lymph node, through a series of lymphatic vessels, and eventually into the blood circulation. In the blood, antibody passes throughout the body and combines with antigen wherever it is found. This antigen-antibody complex is then recognized and destroyed by phagocytic macrophages. The lymphocytes which circulate may settle in a lymph node elsewhere and continue their development there, either transforming into plasma cells or remaining inert as "memory cells" until stimulated to a

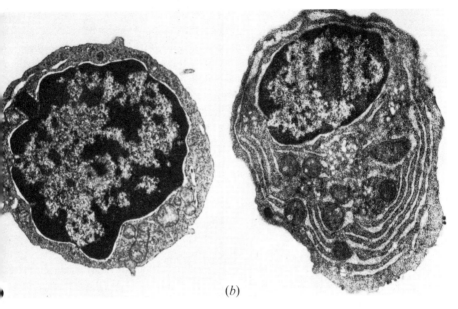

(b)

Figure 9.2 Cells involved in the immune response: (a) a lymphocyte (b) a plasma cell.
(× 10 000) (From Bussard, A. (1973)), L'origine cellulaire des anticorps. *La Recherche* **4**/31: 115–24. (figs 7b, 7c, p. 122).

secondary response by a second encounter with the original sensitizing antigen.

Antigens in solution or suspension, then, evoke antibody which is likewise in solution. If the antigen is an integral component of the cell membrane of an allogeneic tissue graft (that is, one from a genetically different individual) it cannot pass readily to a lymph node and sensitize lymphocytes there. In this case, its presence is detected by a class of patrolling lymphocytes which become sensitized and divide, but do not transform into plasma cells or secrete soluble circulating antibody. They appear to carry antibody molecules on their surface membranes, which form receptor sites specific for the inducing antigen. If a sensitized lymphocyte primed in this way with cell-borne antibody then encounters the allograft, it "recognizes" the antigen and releases a cytotoxic substance which kills the graft cells in the immediate vicinity.

In certain circumstances, such as when a large excess of antigen is employed, or the immune system is still immature, a condition of specific tolerance may develop instead of immunity.

The lymphocytes involved in cell-mediated immunity, and those which take part in the soluble antibody response, represent two separate subgroups of the lymphocyte population with different developmental origins. Both groups originate in the bone-marrow; the cells involved in cell-mediated immunity are those which have passed a critical stage of their maturation in the thymus and are termed T-cells. The remainder are called B-cells: (B stands either for bone-marrow or for the bursa of Fabricius, an organ found in birds through which cells of the soluble antibody response must pass during development. No definite mammalian equivalent of the bursa has been found, although several candidates, such as the liver, have been proposed.) T-cells and B-cells interact and co-operate with each other, and with macrophages, in various subtle and complex ways; for the most part these still have to be elucidated.

One further group of substances is involved in many reactions between antibody and target cells. Collectively, these are known as Complement (C'). Complement consists of 11 protein components, present in the circulation. If antibody molecules attach by their antigen-combining sites to the surface membrane of, for example, a bacterial cell, the components of complement assemble themselves sequentially on the unattached nonspecific ends of the antibody molecules. Once the complex is assembled, part of it has the ability to punch a hole in the cell membrane and cause lysis of the cell. Assembly of the whole system will occur only if a number of antibody molecules have combined with their target antigens, so, although the actual destruction of the target cell is mediated by nonspecific action of complement, the recognition of the cell to be destroyed is highly specific.

The remarkable specificity of the immune response has been exploited in a large number of ways as an experimental tool. One of these, radioimmunoassay of hormone concentrations, has already been mentioned (chapter 1), but many other immunological techniques have been used in the study of reproduction. Immunofluorescence involves the production of specific antibodies which are conjugated with fluorescent dyes. The presence and precise position of the antigen can then be established by allowing combination with the fluorescent antibody in, for example, a thin histological section and observing the position of fluorescence in ultraviolet light. For electron microscopy, an analogous technique has been developed, using antibody conjugated to an electron-dense material such as iron-containing protein. In order to distinguish between the components of a mixture of antigens (such as the proteins of blood or seminal plasma) antibody is raised to the mixture. The separation of the components by diffusion or electrophoresis in a gel medium can then be shown by precipitation with the antibody mixture. Each antigen forms a separate precipitate with its corresponding antibody. These are merely examples selected from a wide range of techniques whose diversity reflects the ingenuity of experimental design as much as the power and sensitivity of the immune system.

9.2 Antigenicity of semen

In copulation, the reproductive tract of the female receives an ejaculate which consists of millions of spermatozoa suspended in a seminal plasma of great complexity (chapter 2). The presence of semen in the female tract is transient, but in most species frequent. If sperm and the accompanying seminal fluid are antigenic, and if an immune response can occur in the female tract, there is at least the possibility of an immunological block to fertilization.

Antigenicity of the spermatozoa in at least some circumstances is readily demonstrated. If semen, or a suspension of sperm, or testis homogenate from a variety of different species is injected into guinea pigs, the blood serum of the recipients acquires the ability to agglutinate and immobilize normal ejaculated spermatozoa (figure 9.3). Care is needed in interpreting

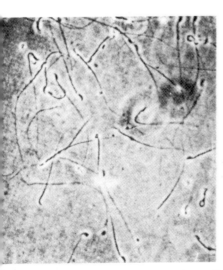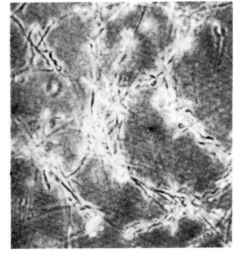

Figure 9.3 Effects of sperm-agglutinating antibodies: *left*, normal sperm suspension, *right*, agglutinated sperm. (From Edwards, R. G. (1967)). Antibodies and fertility, *Science Jnl.*, **3**, 68–73 (p. 70).

experiments of this nature, since the blood serum of some virgin female animals naturally contains a sperm agglutinating factor which cannot be an induced antibody, and is probably a lipoprotein-steroid conjugate. Making allowance for such complicating factors, the ability of sperm to induce spermagglutinin antibodies is now well established.

Partly because of the quest for immunological methods of fertility control, much effort has been devoted to elucidating the nature of spermatozoal antigenicity. A number of different antigens are detectable in sperm. They can be divided into two categories: those which are peculiar to sperm and those which are found also on other tissues.

It is not easy to establish that a particular spermatozoal antigen is endogenous. Any antigen which occurs also in the seminal plasma may well owe its presence on sperm to secondary acquisition from the plasma. Some workers have, however, demonstrated quite clearly the presence of antigens apparently completely absent from seminal fluid, or any other tissues or organs of the body, and have even localized them to specific parts of the spermatozoon. In the guinea pig, for example, four such antigens have been described and designated P, S, T and Z. P is a protein of molecular weight probably somewhat less than 60,000; at any rate, it behaves in sucrose gradient centrifugation and in other respects like albumen of this molecular weight. It is present within the proacrosomal granules and in other parts of the acrosome. S and Z antigens are probably glycoproteins; the former is also found in the acrosome. The chemical nature of T antigen is unknown; it is located on the acrosomal and cytoplasmic membranes of sperm. Several other spermatozoal antigens have been described and located by immunofluorescence techniques on other parts of the spermatozoon, such as the tail, posterior part of the head, and equatorial region (figure 9.4).

Many of the spermatozoal antigens show cross-reactivity between species. Antibody raised against the guinea pig antigen S will combine strongly with the corresponding spermatozoal antigen from rabbits, rats, mice or men; and immunofluorescence techniques have, in bulls, rabbits and humans, shown near identity with respect to one antigen of the equatorial zone of the sperm, and a second on the posterior part of the head. Paradoxically the same antigens which, within an individual male are narrowly restricted to spermatozoa, appear to be widely distributed among different species.

In contrast to these sperm-specific antigens, many of the antigens carried by sperm are common to other tissues as well. Among these are blood-group antigens which occur on erythrocytes, and histocompatibility (H)

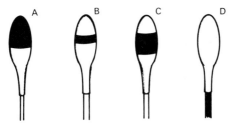

Figure 9.4 The principal patterns of staining of spermatozoa obtained by immunofluorescence. (From Hjort and Hansen (1971)).

antigens, present on most cells of the body. These are the antigens which are primarily responsible for the rejection of tissue grafts.

The presence on human sperm of antigens of the ABO, MNS and P blood group systems has been shown by a variety of techniques. Sperm in suspension together with appropriate erythrocytes and blood-group antiserum may show mixed agglutination with the erythrocytes. This can be due only to the possession of antigens in common, a finding confirmed by the use of immunofluorescence and a battery of other immunological techniques. Still controversial is the question of whether ABO blood-group antigens are present on the sperm of all men, or only on the 76% or so of the (English) population who are "secretors": those who secrete significant amounts of blood group and other tissue antigens in their body fluids. On balance, the evidence appears to favour the presence of at least small amounts of the antigen even on the sperm of non-secretors.

The difference of opinion is an important one. If blood-group antigens are carried only on the spermatozoa of secretors, then the implication is clearly that they are not produced by the spermatozoa, but acquired second-hand from the seminal fluid. If, on the other hand, they are present on the sperm of non-secretors, then they are presumably produced by expression of the spermatozoal genome.

Histocompatibility antigens normally evoke a cellular and not a soluble antibody response. This may have made it less easy to establish their presence on spermatozoa; early results were negative. One approach which successfully demonstrated the existence of a major histocompatibility antigen (H-2) on mouse spermatozoa was to inject a future allograft recipient with a suspension of sperm from a male of the same genotype as the allograft donor. Accelerated rejection of the subsequent graft, in a classic secondary response showed that the injection of sperm had caused presensitization of the recipient. Histocompatibility antigens present on the allograft must also have been present on the sperm. This finding has been confirmed, using other techniques, on the sperm of mice and of men. In both species, the histocompatibility antigens borne by the sperm include those most important in causing rejection: H-2 in the mouse and HL-A in man.

An ingenious combination of immunology and electron microscopy has shown that, in humans, the HL-A antigens are distributed thinly over the head and middle piece of the sperm cells, whereas other antigens are carried on both heads and tails. The amounts of histocompatibility antigen are low; much less than on cells such as lymphocytes from the same animal. This confirms estimates on mice which indicate that the amount of histocompatibility antigen present at the spermatozoal surface is only 10% of that on a lymphocyte. Low antigenicity with respect to histocompatibility antigen may represent a mechanism for minimizing the risk of immune attack in the female.

Unlike blood group antigens in secretors, no histocompatibility antigens have been found in soluble form in the seminal plasma. Their presence on sperm must therefore be due to genetic expression of the germ-cell genome at some stage of spermatogenesis, and not by acquisition from the seminal plasma.

The histocompatibility antigens known as H-Y are, in mammals, found only in cells of males (section 8.2). (Strictly speaking, these are not male-specific antigens, since they occur in the heterogametic sex; the equivalent is found in female birds.) Female rabbits or mice presensitized with spermatozoa subsequently reject grafts of male skin in an accelerated response. This is true of *any* male of the species, not just of skin from a male genetically identical to the sperm donor. In fact, immunization with sperm can cause the accelerated rejection, by a female, even of skin from a male of the same genetic strain. Such a male is genetically identical in all respects other than those consequent on his possession of a Y chromosome, hence the designation of the antigens as H-Y.

H-Y antigens are present on some 60% of spermatozoa. Approximately half of the sperm carry a Y chromosome; the fact that about this proportion manifest a Y-dependent antigen suggests that these are predominantly bearers of a Y chromosome. The proportion which carries the antigen is greater than the proportion expected to bear a Y chromosome, a disparity which may be due to the synthesis of some of the surface membrane of the future spermatozoal cells before the completion of meiosis.

If the X and Y chromosome-bearing sperm can be largely identified by the presence or absence of the H-Y antigen, the significance is two-fold. Firstly, there is the theoretical point that this indicates expression of the haploid genome of the spermatozoon after meiosis. The second (practical) point is that identification of X- and Y-bearing sperm might lead to successful techniques for their physical separation and to the consequent ability to determine the sex of offspring—as important in the production of dairy cattle as it would be for humans (see also section 10.5).

At least one further antigen is known to be acquired by sperm from seminal fluid. This is the so-called Sperm Coating Antigen (SCA), secreted by the seminal vesicles and added to the semen as it passes them during ejaculation. It is found even in the semen of totally azoospermic men. Chemically, SCA is a large iron-containing β-globulin, very similar in electrophoretic mobility and other properties to both serum transferrin and its homologue lactoferrin (found in milk; see section 7.4) but, despite some cross-reaction, immunologically distinguishable from either. Because of these similarities, SCA has been given the name "scaferrin".

The physiological function of this SCA is unclear, but a number of interesting suggestions have been made. One is that adsorption of SCA molecules masks the histocompatibility antigens of the sperm surface, which are of greater importance in eliciting immune rejection and therefore in jeopardizing the survival of the sperm. Both normal and infertile women may have, in their serum and cervical mucus, an antibody which combines with SCA and does not harm the sperm. Nor does immunization against SCA affect fertility in the female rabbit. Why antibody should not reduce fertility in these circumstances is not clear, since cross-reacting anti-lactoferrin antibodies, together with complement, can immobilize human sperm, and sperm-immobilizing antibodies from the sera of infertile women are absorbed by a similar sperm-coating antigen on the post-nuclear cap. A possible explanation is suggested by the finding that human serum has the ability to inactivate one of the components of complement (C_3'). The anti-complementary agent has been identified as serum transferrin. If this property of transferrin is shared by the very similar SCA, then SCA might, by impeding the action of complement, confer protection on the spermatozoa against the immune attack of the female.

The seminal plasma also represents an important fraction of the ejaculate. Many of its components are protein, and therefore antigenic; in fact immunological techniques have been much used in studying the composition of seminal plasma. An immune response against a vital component of the seminal plasma might, therefore, be effective in blocking fertilization. However, since the function of most of the components of the seminal plasma is performed very shortly after insemination, and since in many species very little of the seminal plasma penetrates beyond the vagina, seminal plasma antigens may in general be less vulnerable than spermatozoa to immune attack.

9.3 Immune responsiveness of the female tract and sperm survival

An immune response follows the intravenous injection of either sperm or seminal plasma. Antigenicity of semen when injected in this way would of course be irrelevant unless sensitization could also result from administration inside the female reproductive tract. After intra-uterine immunization of rabbits with homologous sperm, moderately high amounts of spermagglutinin are detectable in blood serum, and somewhat lower levels in the uterine and oviduct secretions.

The induction of antibodies which circulate in the blood and are transferred from the blood serum into secretions of the reproductive tract will presumably occur only if antigen can pass unaltered in the reverse direction to reach the spleen or a regional lymph node. There is no direct evidence of this happening in the uterus, but two pieces of indirect evidence support the idea. One is that by a process termed *persorption* large particles can cross the mucosa of the intestine. Secondly, even people who are deficient in IgA may have circulating antibody against, for example, milk protein. Despite the lack of a local immune response, antibody to a protein antigen may be found in the circulation. This again suggests the passage of intact antigen across a mucosal surface. Possibly a similar process could take place across the uterine mucosa with spermatozoal or seminal antigen.

In addition to the systemic immune response, sperm must contend with the local response carried out by lymphoid cells actually within the mucosa of the reproductive tract. Although predominantly IgA, this locally produced antibody can in certain circumstances include IgM, IgG, and IgE, and might be present in the genital tract secretions in the absence of any antibody detectable systemically. An immunofluorescence technique with antibody directed against the secretory component of IgA (section 9.1) indicates that, in humans at least, the uterus—particularly the cervix— is probably the major area in which the local antibody response occurs, and the vagina probably makes a negligible contribution. In other species this might not apply.

Mucosal surfaces in the bronchi of guinea pigs have been shown to possess the capacity for a purely local cell-mediated immune response.

Although there is little direct evidence that the uterine mucosa can do likewise, this may also be possible.

There is, therefore, experimental evidence that deposition of antigen in the female reproductive tract can elicit both local and systemic immune responses. Consequently it is not surprising to find antibodies against various seminal antigens in the female tract without deliberate experimental induction. Blood group antibodies of the ABO system are found in the cervical mucus; potentially these might incapacitate sperm carrying corresponding blood group antigens. Induced antibody against other antigens may also be present; in humans and cattle, for instance, IgA, IgM and IgG can all be found in samples of uterine fluid. In rare instances, an allergic (IgE) response has been reported. In one such case, a woman was sensitized to a constituent (probably a glycoprotein) of her husband's semen. Sexual intercourse was rapidly followed by violent uterine contractions, a skin reaction, asthma, cardiovascular collapse and unconsciousness. A curious feature was that this occurred at the first intercourse and may have been due to an unfortunate cross-reaction. Apart from this, two other cases have been reported of similar allergic reactions to components of semen (in one case to human semen, and in the other to repeated coital exposure to canine semen). Both women were pregnant at the time the symptoms developed; some resultant modulation of the immune response may have facilitated the sensitization (section 9.5).

In other situations which have been investigated, the presence of antisperm antibodies other than IgE is not closely associated with infertility (see next section). This makes the immune relations between semen and the female even more puzzling. Sperm are antigenic and the female reproductive tract is immunologically responsive. The expected immune response may nevertheless not be detectable; but even if a clear immune response does occur, and antibodies against some component of semen are present, fertility is often not noticeably impaired. How do sperm and other seminal components fail to evoke a major immune response, and avoid its consequences even if it does occur?

A few factors which might contribute to sperm survival in an immunologically hostile environment have already been mentioned. The amount of histocompatibility antigen carried on sperm appears to be low. Spermatozoal antigens may be masked by Sperm Coating Antigen and this SCA might, in addition, impede the action of complement. There is evidence that sperm become coated with IgG obtained from either the seminal plasma or uterine secretions. If this antibody is not itself cytotoxic, it might also effectively mask more important antigens. However, this interpretation of the function of adsorbed IgG may not be correct, since sperm with attached IgG are preferentially phagocytozed. The attachment of uterine IgG is particularly to ageing or dead sperm; it might therefore serve to indicate to phagocytes sperm which require to be removed. Sperm in the oviduct (survivors of the passage through the hostile uterus) mostly lack IgG.

Some form of specific immune tolerance might also enhance the prospects of sperm survival. Where serum antibody is present, this is unlikely. There is however evidence that repeated oral immunization depresses the production of IgG and IgM which results from subsequent immunization by more conventional means. It was suggested that IgA induced locally at the gut mucosal surface might have the function of preventing access of antigen to the more destructive IgG system, and depressing its response to antigen if entry does occur. There is no evidence to indicate whether or not the supposed mechanism occurs in the uterus.

Enzymes capable of inactivating spermatozoal antigens are detectable in the female tract, and it has been proposed that these might operate before sensitization could occur. The enzymes in question are found in lysosomes, so they may merely be associated with the digestion of sperm already phagocytozed. They are, in addition, so numerous and ubiquitous that such a specific role seems improbable.

Seminal plasma of rabbits, injected into cows, is highly antigenic. Attempts to raise antibodies by injection into female rabbits failed completely. The probable explanation for this emerged when a search was made for cross-reactions between human seminal plasma and various other human tissues and secretions. Of 23 "seminal plasma" antigens studied, 7 were found to be immunologically identical with human blood serum proteins, while 10 were detected somewhere in at least one of the women studied. Although the remaining 6 showed no cross-reaction and may therefore have been true seminal-plasma-specific antigens, the results do suggest that women (and female rabbits) may not respond to seminal plasma antigens because they are not, despite original appearances, "foreign". This explanation could not, of course, apply to sperm-specific antigens nor, in the majority of instances, to the histocompatibility or blood group antigens borne by sperm.

In chapter 2 it was mentioned that seminal plasma in many species contains appreciable amounts of various prostaglandins, including those of the E series. Prostaglandins are now known to be involved in regulation of immune responsiveness; in particular, prostaglandin E_1 stimulates the production of cAMP, which in turn tends to depress the immune response.

In one species of bat (*Myotis lucifuga*) a seminal vesicle protein has been identified which depresses leucocyte invasion of the uterus. The seminal plasma may therefore contain a number of constituents which in various subtle ways modulate the immune responsiveness of the female in the interests of sperm survival.

Three general considerations also apply. Firstly, there is no reason to seek a single mechanism which alone explains how the immune attack is evaded. All of the factors discussed may contribute and collectively ensure survival. Secondly, whatever the mechanisms of protection of the sperm, the presence of large numbers of sperm in the ejaculate, and frequently an ejaculate whose volume is large in relation to that of uterine fluid, will bias the situation in favour of the survival of at least some spermatozoa. Finally, the uterine environment is not constant. Figure 9.5 shows the levels of IgG, IgA and the C_3' component of complement in the cervical mucus during 9 normal human menstrual cycles. There is a clear fall in all three at the time of ovulation when fertilization could occur. Even if this decreased concentration is due merely to the increase in volume of mucus at this period of the cycle (see chapter 3) rather than a specific depression of immune responsiveness, the dilution would still make immune destruction of the sperm less likely.

9.4 Immunology and infertility

Some of the spermatozoal antigens identified are restricted to spermatozoa, and do not occur elsewhere in the body of the male. An autoimmune response by the male against his own spermatozoa is therefore possible, and is presumably prevented in most cases by the "blood-testis barrier" (see chapter 2). In a proportion of cases of human male infertility of unknown cause, sperm-agglutinating antibody has been detected in the serum. In one investigation of men attending an infertility clinic, 4·9% of those showing azoospermia or oligospermia had detectable spermagglutinin. (Azoospermia denotes a total lack of spermatozoa in the semen, and oligospermia a sperm count of less than 4×10^7 per ml). None of the control group (the husbands of pregnant women) carried serum spermagglutinin. Estimates in other surveys range from 5–12% of infertile men.

Attempts deliberately to immunize against sperm have been carried out in a number of species, of which the guinea pig has proved the most suitable. Sixty days after injection of homogenized testis material suspended in adjuvant (adjuvants are substances which potentiate an

K

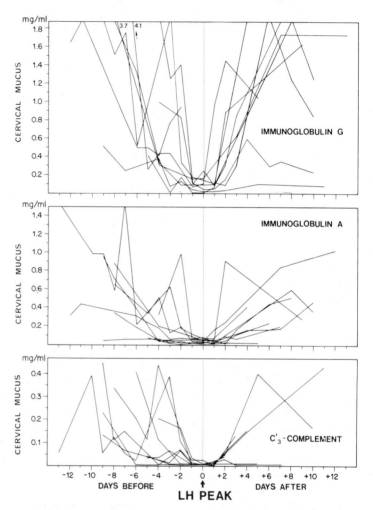

Figure 9.5 Serum levels of IgG, IgA, and of one component of Complement measured in the cervical mucus of women around the time of the mid-cycle LH peak. (From contribution by Schumacher, G. F. B. to discussion of paper by Vaerman, J. P. and Férin, J. (1974)), Local immunological response in the vagina, cervix and endometrium, pp. 281–305 in Diczfalusy, E. (ed.): Immunological approaches to fertility control. Transact. 7th Karolinska Symposium, Geneva, 1974. *Acta endocrinol.* (Kbh.), Vol. 78, Suppl. No. 194, pp. 302–305 (1975). (Karolinska Institutet, Stockholm) (fig A, p. 303).

immune response) synergistic cell-mediated and soluble antibody responses result in a virtually total depletion of the germ cell population of the testis. Similar results have been obtained on human volunteers (figure 9.6); ultimately, some form of immunological contraception may prove practicable.

In women, too, there is some relation between infertility and evidence of an immune response against sperm. The diversity of methods of assay results in estimates of the proportion of infertile women with circulating antisperm antibody varying from 7% to 79% (see Table 9.2). Nevertheless, women who are known to be infertile are more likely to have antibody than women who are not. The figures referring to prostitutes show a relatively high incidence of antibody, presumably reflecting relatively high exposure to seminal antigens. Obviously anomalous are the figures for pregnant

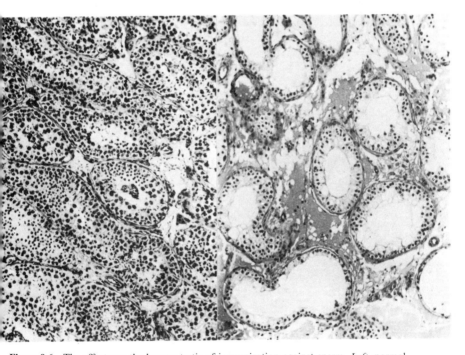

Figure 9.6 The effects on the human testis of immunization against sperm. *Left*, normal seminiferous tubules in transverse section. *Right*, a similar view of a testis after immunization. Note the virtual absence of germ cells within the tubules. (From Mancini, R. E. (1971)), Immunological approaches to fertility control, pp. 157–78 in Diczfalusy, E. and Borell, U. (eds.): Control of Human Fertility. (Wiley Interscience and Almqvist and Wiksell Förlag AB, Stockholm) fig. 3 (p. 166) and fig. 4 (p. 167).

Table 9.2 Percentage incidence of circulating antibody against spermatozoal antigen in the serum of women, assessed by various techniques

	1	2	3	4	5	6	7
Infertile women	14·1	45	70	26	60	19	7
Single women/controls		20		6	11	0	
Pregnant women	0·0	—		26	54	0	
Prostitutes	17·4	72	—				
Nuns	4·2	—					

1. Kolodny *et al.* (1971)
2. Schwimmer *et al.* (1967)
3. Franklin and Dukes (pooled)
4, 5, 6. Isojima (1969)
7. Israelstam quoted in Kolodny *et al.* (1971)

women, and for nuns. In the latter case, the explanation might lie with an autoimmune response against some cross-reacting antigenic component of the female reproductive system, or to residual antibody induced by exposure to seminal antigens before the women in question became nuns.

9.5 Immunological problems of pregnancy

The immune status of the fetus in relation to its mother is very different from that of spermatozoa. Whereas the contact between sperm and the female is a matter of periodic irruptions of short-lived cells, almost all of which must inevitably be destroyed within a short period, the fetus must survive prolonged contact with the maternal endometrium, and a relationship sufficiently intimate to permit reciprocal transfer of nutrients and waste materials. It is in a position much more analogous to that of a conventional tissue allograft, and yet is not rejected during a gestation period possibly lasting many months. How does the fetal allograft survive?

No question of immune destruction of the fetus would arise if the fetus was not antigenic or, at least, if expression of histocompatibility antigens specified by the paternal part of its genome was delayed until birth. The need to postulate some mechanism which might make this feasible has been avoided by the clear demonstration of histocompatibility antigens on mouse embryos even before implantation. The techniques used were much the same as those which demonstrated the presence of histocompatibility antigens on spermatozoa (see pp. 139–40).

It has also been suggested that the ability of the mother to respond to allografts is generally reduced during pregnancy. Skin allografts do last somewhat longer on pregnant animals; in rabbits the rejection time is almost doubled if the recipients are 3 weeks pregnant at the time of grafting. In mice, on the other hand, no prolongation is detectable. "Immunological inertia" consequent on pregnancy does not seem an adequate general explanation of survival, even of the first brood, in species with a gestation period much greater than 2 or 3 weeks.

The immunological inertia of pregnant rabbits was originally attributed to high circulating levels of corticosterone during pregnancy. Experimental treatment with corticosterone or ACTH in mice prolongs graft survival by 50%. Other hormones present at elevated levels during pregnancy have similar effects; oestradiol, progesterone, and simultaneous injection of HCG with cortisone have all been found, in various species, to have some effect. General immune depression by elevated levels of any or all of these hormones in the circulation may not explain fetal survival; however, it is possible that local secretion of hormones by the conceptus might locally impede some aspect of the immune response. Apart from steroid secretion by the trophoblast (chapter 6), the oestrogen and HCG-like hormone carried by some pre-implantation blastocysts (p. 86) might be important in the early stages of gestation. It also seems probable that, in humans, placental HCG and HPL (section 6.1.3) have the effect of blocking the antigen receptor sites of maternal lymphocytes. Prostaglandins may suppress lymphocyte activation; the blastocyst of the rabbit has been shown to be capable of synthesizing prostaglandins. Finally, a protein produced by the fetus and present in the amniotic fluid (α-fetoprotein, AFP) can suppress lymphocyte activation *in vitro* at concentrations as low as 1 μg ml^{-1}. The significance of these mechanisms is hard to assess, but individually or collectively, they contribute to fetal survival.

A further possibility is that the uterus may be a privileged site for graft survival. This would be the case if, for example, it lacked sufficient afferent lymphatic drainage by which antigen could reach regional lymph nodes and effect sensitization. (The efferent arm of the immune reflex, the blood supply which would carry antibody or sensitized lymphocytes back to the area of the graft, obviously cannot be deficient.) On histological evidence, this suggestion seems unlikely, but two experimental findings seem finally to rule it out. Firstly, as we have seen, antibody can be induced by intrauterine immunization (p. 142). Secondly, if the immune response is blocked by a lack of lymphatic drainage of the uterus, the block would be circumvented by injecting paternal antigen directly into the circulation, or grafting paternal tissue to some part of the body outside the putative privileged site. Immune rejection of the fetus should ensue; this does not happen. Hence it can be concluded that the uterus is not in this sense a privileged site for the allograft survival.

Survival of the fetus would be accounted for if there were some form of barrier which could prevent either antigens leaving the conceptus, or maternal antibody or lymphocytes entering it. A likely candidate for this barrier is the trophoblast (chapter 5) which, in mice, forms a virtually complete layer between an established conceptus and the endometrial cells of its mother. Trophoblast appears to be only very weakly antigenic. A mouse blastocyst transplanted ectopically (that is, to a site other than its normal one) with its trophoblast layer intact will thrive in many situations, such as within the kidney capsule, or even inside the testis capsule of a male rat: the wrong site, the wrong species and the wrong sex! Embryonic tissues transplanted without surrounding trophoblast are destroyed in an immune response.

The low antigenicity of trophoblast appears to be explained by a thin covering layer of fibrinoid sialomucin (figure 9.7). Treatment with neuraminidase removes this fibrinoid layer and renders the trophoblast capable of evoking an immune response. Why sialomucin should have this property is not clear. A simple masking of antigenic sites may be the explanation. Alternatively, it has been suggested that it confers on the trophoblast a relatively high negative surface charge which reduces contact between trophoblast and the patrolling lymphocytes, which likewise are highly negatively charged. This electrostatic repulsion may prevent sufficiently close contact for the lymphocytes to become sensitized.

Recent electron-microscope studies cast doubt on this suggestion. In a number of species, maternal lymphocytes, decidual cells, and trophoblast are very closely intermingled, while rabbit placentae may totally lack fibrinoid, and even in mice the fibrinoid is lacking in early gestation and discontinuous in late pregnancy. The concept of a trophoblast barrier based on fibrinoid coating is not an adequate explanation.

Perhaps more conclusive evidence against the existence of an absolute barrier is the finding that the mother does become sensitized to the fetus. One of the early signs of sensitization to an allograft is hypertrophy of the relevant regional lymph nodes. The para-aortic node, which receives lymphatic drainage from the uterus, hypertrophies if the uterus contains either an allograft or an antigenically foreign fetus. Antibody specific to major histocompatibility antigens has also been detected in humans and mice, while experimentally-induced anti-fetal antibody crosses the placenta and enters the fetus.

Similarly, fetal cells may cross the placenta and sensitize the mother. A

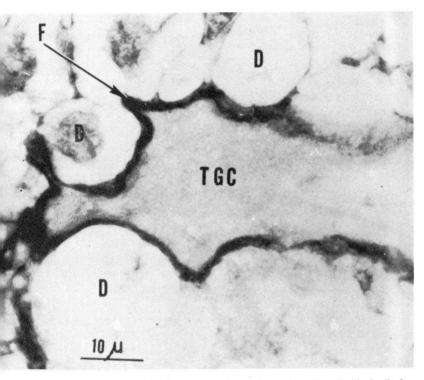

Figure 9.7 Electron micrograph of the maternal-placental boundary. D = decidual cell of uterus, TGC = giant cell of trophoblast, F = fibrinoid sialomucin layer. (From Bradbury *et al.* (1965)): A histochemical and electron microscopical study of the fibrinoid of the mouse placenta. *J. Roy. Microscop. Soc.* **84**/2: 199–211 (fig. 1B. p. 202).

special case of this is the cause of Rhesus haemolytic anaemias caused by sensitization of a mother who lacks the Rhesus antigens to the red blood cells of her Rhesus-positive fetus. This will probably not affect the fetus which induced it; subsequent Rhesus-positive fetuses are at risk of severe anaemia or death resulting from maternal antibody crossing the placenta and destroying fetal red blood cells. The Rhesus syndrome does not arise where fetal red blood cells are also incompatible with respect to ABO blood group antigens, since in this case the corresponding natural antibody in the maternal circulation will react with the fetal cells as soon as they enter. There is no opportunity for sensitization to occur. An effective means of preventing the Rhesus syndrome involves similar removal of Rhesus-positive red blood cells by supplying the Rhesus-negative mother with exogenous anti-Rhesus antibody. Since this antibody is also antigenic, it disappears from the circulation before it could itself harm a subsequent fetus.

Other fetal cells besides red blood cells can enter the maternal circulation and, conversely, small numbers of maternal lymphocytes may cross the placenta in the reverse direction. The weight of experimental evidence does not, therefore, suggest sequestration of the fetus from the maternal immune system. On the contrary, the immune system of the mother does recognize and respond to the fetus but, despite the presence of antifetal antibody and sensitized lymphocytes, and the passage of both into the fetus, the response is rarely a destructive one.

A possible explanation is that the soluble antibody combines with its corresponding antigenic sites on fetal cells, but is not itself cytotoxic. This might be due to sparseness of antigenic sites on the fetal cells, or perhaps to a lack of complement (or one of its components), or to an anti-complementary agent in the fetal tissues. Once soluble antibody has combined with histocompatibility antigens, they would no longer be available for recognition by the invading maternal lymphocytes. The phenomenon is known as "specific immune enhancement" from its discovery in a series of experiments which were intended to suppress tumours by prior immunization to tumour-specific antigens, but succeeded only in enhancing their growth.

A further piece of evidence circumstantially supporting some such fetal manipulation of the maternal immune system is the finding that in early human pregnancy the number of circulating T cells declines relative to B cells. This might conceivably reflect a depression of the capacity for a cellular immune response and consequent encouragement of soluble antibody formation, circumstances presumably conducive to the production of a state of specific immune enhancement. If enhancement or some similar mechanism is involved, the fetus survives by playing one half of the immune system off against the other.

Some form of manipulation by the fetus of the immune system of its mother certainly seems to occur. Apart from the possibility of immune enhancement, immunosuppressive effects of α-fetoprotein, fetal lymphocytes, and trophoblast hormones have all been reported. Trophoblast, although an imperfect barrier, may serve a useful ancillary function by reducing contact between fetus and mother. As with the immunological survival of spermatozoa, so with the fetus; there is no need to invoke one single mechanism. Several different factors may each contribute to survival of the fetal allograft.

9.6 The passive acquisition of immunity by fetus and newborn

The immune system of a mammal does not become functional until around the time of birth. Before birth there is little need for a fetus to combat infection. Immediately after birth, the maturing immune system is exposed to a range of pathogens of which it has had no experience and against which, in consequence, it has not yet produced antibody. This vulnerable period is made less hazardous by the transfer of antibodies from the mother to her young.

Antibody may be acquired either before birth, by absorbing it through the yolk sac or across the placenta, or after birth in the milk, or both. The means of acquisition of a number of species are summarized in Table 9.3. In many species, antibody is acquired via the yolk sac, probably as a cross-section of that currently circulating in the mother; it includes both IgG and IgM. Transfer is not by simple diffusion, since not all proteins of equal molecular size are equally transported. In rabbits, rabbit IgG passes into the fetus at approximately 100 times the rate of injected bovine IgG. In primates, transfer is placental. As IgM cannot cross the placenta, the antibody acquired is principally IgG. This excludes the majority of ABO blood group antibody and therefore permits successful ABO-incompatible pregnancies. This may be why it is in primates that blood group systems with naturally-occurring antibody are found.

Postnatally antibody transfer may occur by secretion into milk (especially colostrum) and absorption from the gut. Immunoglobulins, being protein, risk being digested in the duodenum of the·newborn as

Table 9.3 Transmission of passive immunity.

	prenatal	route	postnatal	time (days)
horse	−		+ + +	1
pig	−		+ + +	1–1½
ox, goat, sheep	−		+ + +	1
wallaby	−		+ + +	180
dog, cat	−	?	+ +	1–2
hedgehog	+	?	+ +	40
mouse	+	Y?	+ +	16
rat	+	Y	+ +	20
guinea pig	+ + +	Y	−	
rabbit	+ + +	Y	−	
man, monkey	+ + +	P	−	

Y = yolk sac
P = placenta

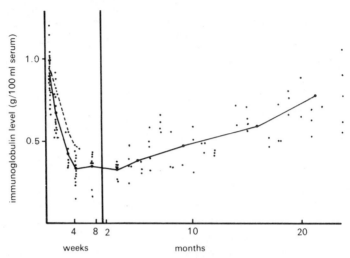

Figure 9.8   Changes in serum levels of immunoglobulins from birth to two years in the human. (N.B. alteration in time scale). (From Brambell (1970)).

trypsin secretion starts. To avoid this, and ensure safe passage of antibody molecules to the absorptive regions of the small intestine, a potent trypsin inhibitor is also secreted in the milk. In pigs, 1 ml colostrum can inhibit as much as 2 mg trypsin. Experimental administration of trypsin inhibitor increases immunoglobulin uptake.

The period over which antibody uptake from the milk can occur varies according to species. In the rat, the gut becomes impermeable 22 days after birth, although antibody secretion into the milk continues.

Primates show no significant absorption of antibodies from the gut into the circulation. Nevertheless, IgG, IgA and IgM antibodies are secreted in the colostrum and early milk. These are primarily directed against gut organisms and are presumably effective in the gut rather than in the circulation. Figure 9.8 shows the antibody levels in the blood of human babies from birth to 2 years of age. The initial decline represents loss of prenatally-acquired antibodies (the half-life of which is around 3 weeks) followed by the progressive accumulation of endogenous antibody. No contribution is made by antibodies from the milk.

CHAPTER TEN

REPRODUCTION AND SOCIETY

10.1 Reproduction and society

Little is of such fundamental concern to us as sexuality and reproduction. Human reproduction, its mechanisms and the factors which influence it, are matters of more than academic interest. This has always been the case, but at a time in history when the human population has been increasing at a greater than exponential rate, it has become more important than ever before to understand the processes of human reproduction in order to control them.

One objective of reproductive research is therefore the achievement of control of means of reproduction. This chapter discusses a number of fields in which artificial control of reproduction. has been exerted. It is possible to select only a few examples: unwitting social effects on reproduction, deliberate contraception, preselection of fetal sex and prevention of the birth of congenitally malformed infants, and future possibilities such as the genetic manipulation of embryos.

10.2 Changes in the pattern of human reproduction

It is well known that the human population has been increasing with frightening rapidity, and will continue to do so for some time to come. The causes of this rise are complex, but it seems reasonable to view the modern rise in population (in western Europe, at least) as a transition from a population with high fecundity offset by high mortality, to one with low mortality and low fecundity. The decline in reproductive rates seems to follow declining mortality only after a delay; it is this excess of births over deaths which is largely responsible for population increase. The existence of the higher equilibrium—or, at least, its maintenance by low birth and death rates—is somewhat speculative; if the birth rate fails to decline, mortality will inevitably rise again.

Until relatively recently, mortality in Europe was highest among infants and young children. From a demographic point of view, this is particularly important, since high mortality during the pre-reproductive years has much more relevance to the rate of population increase than mortality after the age at which further reproduction becomes unlikely. Figure 10.1 contrasts typical patterns of mortality in human populations.

A second important factor influencing the rate of increase of a population is the age at which people start to reproduce. Social influence may delay procreation for some time after puberty, in humans as in many other animals. However, the age of puberty will be an important determinant of fertility.

The interval between successive births will also be important in determining reproductive rates. Lactation suppresses ovulation and therefore has some contraceptive value. In the absence of any effective means of artificial contraception, breastfeeding will postpone ovulation and extend the period intervening before a second conception. Estimates of the effectiveness of breastfeeding vary. In one survey (of Rwandan women) half of a sample of breastfeeding women had conceived again within 18 months; in a group of women who did not breast feed, the corresponding figure was 3 months.

Infant mortality, the age of sexual maturity, and lactational inhibition of ovulation (lactational amenorrhoea) probably played a significant part in controlling human population growth in non-industrial societies. As Roger Short has pointed out, the effectiveness of all three as means of population

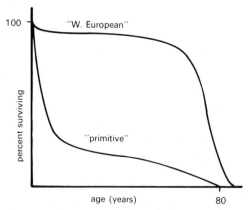

Figure 10.1 Survival curves for hypothetical human populations: (*a*) the "primitive" pattern, characterized by high infant mortality; (*b*) the "advanced" population, with lower infant mortality.

control has been vitiated in recent times. Infant mortality has sharply decreased and consequently a higher proportion of those born survive to reproductive age. The immediate reasons for the fall in mortality are diverse, and not within the scope of this book, but there are good reasons for implicating improvements in nutrition and public health as being underlying causes of great significance.

The onset of menstruation (menarche) is a readily observed sign of sexual maturity, hence detailed information on the age at maturity is available for girls. In western societies, there has been a steady and consistent decline since records were kept. There is evidence that this trend has now ceased, in some countries at least. Data earlier than the middle of last century, and data on the age of sexual maturity in boys, are not available.

Puberty in girls is more closely related to body weight than to age (section 8.5). Improved nutrition would therefore result in faster growth in childhood, and the early achievement of the body weight threshold critical to the start of menstruation. It is likely that decline in the age at menarche is in large part due to a rise in standards of nutrition and general health.

Breastfeeding for prolonged periods is less common in industrial than in non-industrial societies. An analysis of books on child rearing from the fifteenth to the twentieth century indicates that the period of breastfeeding accepted as normal (at least in the eyes of the authors) declined greatly in the eighteenth century. A decline in breastfeeding therefore appears to coincide with the start of a period of rapid population growth and (insofar as breastfeeding has a contraceptive effect) must have contributed to it. The significance relative to other factors is hard to assess, but it has been estimated that, in the Philippines, the number of births would be 20–25% higher were it not for the contraceptive effect of breastfeeding.

A social factor—the decline in the length of time for which mothers breastfeed their infants—thus would tend to increase the likelihood of conception closely following birth. Additionally, breastfeeding appears to have lost some of its effectiveness in postponing the return of fertility. In part, this may be due to nutritional improvement. Lactational suppression of ovulation is a device for preventing the excessive metabolic demands of pregnancy simultaneously with lactation. It is reasonable to conjecture that improving standards of nutrition would allow earlier ovulation than otherwise would be the case. Another possible explanation is that techniques of breastfeeding have altered. In a primitive society, the infant is likely to have much more prolonged contact with the nipple. Consequently, the ovulation-suppressing effects of suckling may be more intense.

Changes which are associated with the transition from a rural to an industrial society therefore act to reduce perinatal mortality, to accelerate sexual maturity, and to reduce the contraceptive efficacy of breastfeeding. The result was a dramatic, and continuing, rise in population. There is no question of an indefinite increase in population: the only point at issue is whether the human population will be controlled by famine and pestilence, after outstripping available resources, or whether means will be found to reduce the rate of increase by more acceptable means. Social control such as delaying the age of marriage may be important, but in conjunction with artificial control of fertility. The next section deals with current means of contraception.

10.3 Contraception

10.3.1 *Contraception in women*

In principle, interference with almost any part of the complex female reproductive system could significantly impair fertility. In practice the most efficient contraception (surgical methods apart) has resulted from manipulation of the interplay of hormones controlling ovulation, in particular by various forms of orally administered steroids. Many different formulations are now available, but most consist of combinations of oestrogens and progestogens, administered either simultaneously or sequentially. Figure 10.2 gives the chemical structure of a variety of steroids used in modern oral contraceptives.

The primary action of most steroid contraceptives is to supress ovulation by inhibiting the ovulatory surge of LH. Administration of LRF to women using combined-type steroid contraception is followed by a rise in plasma LH. This suggests that the major action of the contraceptive steroids is on the hypothalamus, and that they inhibit LRF release, but an effect on the pituitary is also possible. At any rate, during a pill "cycle", FSH and LH both remain at low levels. In addition, ovarian sensitivity to gonado-trophins is reduced. A few ovulatory cycles may occur (at a frequency of around 15–20 per 100,000 with combined-type oral contraceptives and perhaps ten times this number with sequential types), but the chances of these resulting in conception is extremely slight, since steroids act also at other levels of the reproductive system. In a normal cycle, oestrogen and progesterone have a wide range of effects which facilitate conception, and many of these steroid-dependent functions are impaired by synthetic steroids. Motility and metabolism of the oviduct and uterus are altered; the

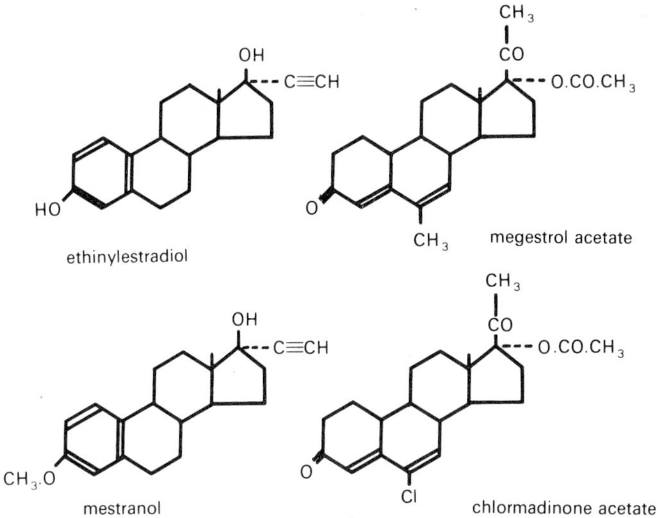

Figure 10.2 Examples of steroids synthesized for use in the formulation of oral contraceptives. Ethinylestradiol and mestranol are oestrogens, megestrol and chlormadinone progestogens. Cf. Fig. 4.6.

entry of the egg into the uterus may therefore be mistimed, and the receptivity of the uterus reduced. The consistency of cervical mucus in a normal cycle becomes transiently favourable for sperm passage in response to the ovulatory rise in oestrogen; under the effects of contraceptive steroids, this does not occur, and the sperm cannot enter the uterus. Finally, there is some evidence for impairment of sperm capacitation.

As steroids (particularly oestrogens) have numerous metabolic effects, it is not surprising that various side-effects of steroid contraception have been reported. The most serious of these are circulatory: high blood pressure and thromboembolism (the formation of potentially fatal clots in blood vessels). Nevertheless, taking into account the risks consequent on pregnancy, and relative effectiveness of steroids, oral contraceptives compare favourably with the alternatives (Table 10.1).

There are in addition directly beneficial effects, such as a reduction in the frequency of certain forms of breast cancer. It has been suggested that the relatively high incidence of breast cancer in western society results from the biologically unnatural occurrence of repeated menstrual cycles. Until recently, a woman would spend much of her adulthood lactating or pregnant. Exposure of the breast and other steroid target organs to the

Table 10.1 Effectiveness of contraceptive methods, and estimates of mortality resulting from their use.

Method of control	Estimated method failure per 100 woman-years	Estimated number of deaths from method	Total deaths from pregnancy and method
A. Developed countries			
None	—	0	15
Oral contraceptives	1·0	3	4
IUD	3·0	1	2
Condom and diaphragm	20·0	0	4
B. Developing countries			
None	—	0	300
Oral contraceptives	2·0	3	13
IUD	3·0	1	16
Condom and diapharagm	20·0	0	75

Figures from *Population Reports A2:35*. Calculations are based on the assumption of death rates during pregnancy of 25/100,000 births in developed countries and 500/100,000 in developing countries. Death rates are annual, per 100,000 women aged 15–44.

frequent fluctuations in level of circulating steroids is a recent phenomenon.

Many of the harmful side-effects of contraceptive steroids appear to be associated with the oestrogen component. This has led to the development of progestogen-only oral contraceptives. Unlike combined types, these suppress ovulation only in some cycles (15–40%, but evidence is scant). Much of the contraceptive effect comes from secondary effects on motility of the oviduct, on luteal function, on the cervical mucus, and on endometrial receptivity, together with a disruption of the timing of the LH rises. The disadvantages of progestogen-only methods lies in their lower effectiveness and a higher incidence of ectopic implantations among the conceptions resulting from method failure.

The other widely-used form of female contraception is the intra-uterine device (IUD). The presence of these foreign bodies in the uterus can affect reproduction in a number of ways: alterations in endometrial histology, leucocyte infiltration of the uterus resulting in phagocytosis of sperm or eggs, luteolysis, altered motility of the female tract, or the release of cytotoxic substances into the lumen. The effectiveness of IUDs can be enhanced by designing them so as to maximize the area in direct contact with the endometrium, or by preparing them in such a way that they release small amounts of copper or progesterone into the lumen over a period of time. A copper-releasing IUD inhibits sperm motility *in vitro*, but apparently not *in vivo*. Both copper-and progesterone-releasing IUDs appear to act by altering the uterine environment.

IUDs are second only to oral steroid contraceptives in effectiveness. Net pregnancy rates range form 0·0 to 5·6 per hundred women for the first year after insertion, compared with rates of 1–3 per hundred women-years for oral contraceptives.

Much interest has recently been shown in the development of further forms of contraception. Amongst these are the use of synthetic LRF analogues and immunization against specific hormones such as HCG, as well as more ingenious suggestions such as specific inhibition of capacitation, or acrosin action (section 3.3). These have in some cases been shown to work on other animals, but are not yet fully ready for human use.

10.3.2 *Contraception in men*

Compared with the inhibition of ovulation, prevention of spermatogenesis presents a less tractable problem. In women, a single monthly ovulation must be suppressed or, if it occurs, its synchrony with uterine events can be disrupted. In males, spermatogenesis is continuous and vast numbers of sperm are normally produced.

Nevertheless, a similar approach is feasible. As in the female, gonadal steroids act on the hypothalamus to exert negative feedback control of gonadotrophin release. In principle, artificial administration of steroids will therefore suppress gonadotrophin production and therefore spermatogenesis. Testosterone itself is effective, in the form of testosterone propionate. Daily intra-muscular injections of 25 mg testosterone propionate for 60 days progressively lowers the sperm concentration of semen, culminating in total azoospermia. The effect is totally reversible; 150 days after the last injection, semen contains normal concentrations of fully motile sperm.

The drawback to this form of contraception is the need for daily, and painful, injections. Less frequent and larger doses are unreliable as a form of contraception. Many orally active androgens have also been tried, but these have so far been ineffective. If given in doses high enough to reduce sperm production effectively, they also reduce libido and potency; in suppressing pituitary gonadotrophin release they reduce testosterone output by the Leydig cells, and are themselves too weakly androgenic to compensate.

L

Oestrogens and progesterone inhibit gonadotrophin release in men, as in women. Current strategies in developing male contraception have centred on exploiting this, while simultaneously administering androgen to eliminate the concomitant loss of libido. One successful combination is a combination of ethinylestradiol (figure 10.2) with methyltestosterone, an active androgen with a longer half-life than testosterone itself. In one clinical trial, all of the subjects became totally azoospermic after 18 weeks of treatment. Sperm reappeared in the semen 100 days after the treatment was stopped, and spermatogenesis appeared to be normal again by 35–40 weeks. Side-effects were mild; although some of the subjects reported temporary loss of libido, this was during a period of placebo administration as well as in the early stages of steroid treatment. There seems no good reason, therefore, why an effective "male pill" should not eventually be developed.

Other methods of chemical contraception in men are being investigated. Among these are prolonged steroid administration by implanting silastic capsules from which steroids are released over a period of time, and attempts to interfere locally with the function of the epididymis and maturation processes which the sperm undergo there.

Apart from physical barriers to fertilization, vasectomy is the only widespread form of male contraception currently practised. Compared with the corresponding operation in women, it is simple and quick to perform and, if sufficient time elapses to allow loss of all sperm distal to the cut, it is completely reliable. Side-effects are few; occasional haematomata and inflammation are the commonest, affecting less than 3% of cases. Frequently, vasectomy is followed by a rise in circulating levels of anti-sperm antibodies. These may help to eliminate excess sperm in the blocked genital tract. They do no apparent harm, although obviously might delay or even prevent return to full fertility after reversal of vasectomy. Deliberate immunization against sperm has also been shown to reduce fertility in men.

10.4 Abortion

If conception occurs, birth can still be artificially prevented. Three main techniques are used: minor surgical procedures, infusion of hypertonic saline into the amniotic sac, and the application of stimulants of smooth-muscle contraction, such as prostaglandins.

Curettage and vacuum aspiration of the contents of the uterus are most suitable for terminating early pregnancies. During the first three months of pregnancy, the operation is extremely safe: maternal mortality according to one survey is only a fraction of that recorded for other minor operations such as tonsillectomy. Risk to the mother rises sharply later in pregnancy, but even so a mother is less likely to die as a result of an abortion during the second three months than she is if the pregnancy is allowed to continue to term without intervention.

The mode of action of hypertonic saline is not clear. It appears to increase the uterine production of prostaglandins, but may in addition suppress placental progesterone synthesis and increase the release of oxytocin. All three effects would facilitate expulsion of the conceptus (section 7.2).

Prostaglandins E_2 and $F_{2\alpha}$ have become extensively used as abortifacients, administered by injection, intrauterine infusion, or intravaginally. Their action mimics normal labour and prostaglandins have been used in the artificial induction of labour at term. An alternative use of prostaglandins is in "menstrual regulation": the artificial (or premature) removal of the decidualized endometrium, some time after ovulation, whether conception is known to have occurred or not.

Apart from the use of abortion as a means of birth control, it can also be used for what may be broadly termed eugenic purposes. If a developing fetus is identified as grossly abnormal or defective, the decision can be made to prevent its birth.

Many prenatal screening methods are now available, or soon will be. Major morphological defects, such as anencephaly can be detected by ultrasound scanning of the abdomen of the mother. Anencephaly and spina bifida are associated with a rise in the level of α-fetoprotein (AFP) (sections 8.4 and 9.5) in amniotic fluid. This can also be detected in maternal blood serum as early as 16 weeks of gestation. It is now possible to obtain and culture a sample of fetal cells from the amniotic fluid. This technique, amniocentesis, makes it possible to decide whether the chromosome constitution and metabolism of the cells is normal, hence whether the fetus is carrying serious genetic disorders. Among the disorders which can be identified by these means are Down's syndrome ("mongolism") and Tay-Sachs disease (muscular weakness, convulsions, blindness and mental retardation appearing in children of about 10 months).

Induced abortion must be seen in the perspective of normal fetal losses. Estimates vary, but it seems unlikely that less than about 25–30% of all human conceptions abort spontaneously. The figure may in fact be very much higher than this, but there is practically no way of getting precise information about embryonic mortality within the first few days after conception. Whatever the method of spontaneous abortion, it appears to be selective. The proportion of spontaneously aborted embryos with chromosome abnormalities is higher than that among successful pregnancies. Screening for defects and selective abortion appears to be one of the routine functions of the uterus. It has even been postulated that certain apparently teratogenic agents are in fact saving the lives of abnormal fetuses which would otherwise have been recognized as such and spontaneously aborted. One such possible example is the association between potato blight and a high incidence of spina bifida and anencephaly; laboratory experiments have failed to show a directly teratogenic action of potato blight.

10.5 Preselection of sex

At birth, the sex ratio (defined as the ratio of males to females) in most mammals is not far from unity. Many people would wish it otherwise; among them cattlebreeders and prospective parents. Apart from selective infanticide (which has been extensively practised in various human societies) no effective methods have yet been devised for controlling sex ratio.

Determination of fetal sex is possible by amniocentesis and chromosome analysis of the cells obtained. This is facilitated by the recent use of the dye quinacrine, which causes an intense fluorescence of Y-chromosomes, and consequently permits the ready identification of male cells. Prenatal sexing of a fetus could then be followed by selective abortion of fetuses of the unwanted sex. Although for a number of reasons this approach might not be thought suitable for humans, it could prove to be of value in stock breeding.

A more generally satisfactory approach is to separate sperm carrying an X-chromosome from those with a Y, the former being female and the latter male-determining. The X-chromosome is larger than the Y; hence male-determining sperm are fractionally lighter, and differ in density. Centrifugation of human sperm in density gradients does in fact result in fractions which differ in the proportion of Y-sperm, but so far complete separation of viable X and Y-sperm has not been achieved. Male-determining sperm appear to migrate faster through cervical mucus, but the only claim to have separated sperm by exploiting motility differences was not subsequently confirmed.

It has also been proposed that differences in vaginal pH affect the sex ratio of the resultant offspring, alkaline pH favouring X-sperm. As the pH after coitus is determined largely by the semen and not by vaginal secretions (section 3.1), recommendations such as vaginal douches with baking soda or dilute vinegar before intercourse are probably ineffective.

The timing of intercourse in relation to ovulation is also supposed to affect sex ratio. In a study of 875 pregnancies among users of the rhythm method of contraception, it was found that the longer the interval between coitus and the slight rise in basal body temperature which accompanies ovulation, the greater the proportion of males subsequently born. With artificial insemination, the reverse was true. In neither instance, however, was the proportion of males greater than 68% (i.e. sex ratio = 2·15).

Numerous other methods for predetermining sex have been investigated, ranging from further sophisticated attempts at sperm separation to keeping one's boots on during intercourse (which produces males). None seems to have produced a sex ratio deviating sufficiently from the normal to be of practical value.

10.6 The artificial improvement of fertility

While much research on reproduction is directed towards finding satisfactory means of curbing fertility, some is devoted to increasing it. This is valuable to breeders of domestic animals; it is also important in alleviating human infertility. It has been estimated that more than 12% of couples of childbearing age are unable to conceive. In perhaps half of these cases where the cause of infertility can be diagnosed, effective treatment is possible.

Female infertility can be due to many causes. Among these are failure of the cervical mucus to alter in consistency at the time of ovulation so as to allow passage of the sperm, failure of ova to mature and ovulate, failure to produce sufficient progesterone to prepare the uterus for implantation, asynchrony of any of these events, or a simple physical blockage of the oviducts preventing gamete transport.

Artificial insemination (see below) can be used to bypass inhospitable cervical mucus, and an obstructed oviduct can in some cases be surgically cleared. For the rest, manipulation of the endocrine mechanisms controlling reproduction is necessary.

If ovulation fails, several forms of hormone therapy are possible. One of these involves treatment with clomiphene, which elicits a rise in the serum levels of both FSH and LH and, with luck, ovulation. The mode of action of clomiphene is not clear. It is both weakly oestrogenic and antioestrogenic. Two isomers exist, the cis- and trans-forms (figure 10.3), which are both effective in inducing ovulation. For some reason, pregnancy is much more frequent after treatment with the cis-form. As most clinical treatments so far have used mixtures of both forms, and as the criteria for the occurrence of ovulation are often imprecise, it is difficult to assess the potential effectiveness of clomiphene treatment.

An alternative treatment involves the use of gonadotrophins. As discussed in section 4.3, many oocytes start development but (in humans) all but one of each batch become atretic. Gonadotrophin treatment maintains the development of oocytes which would otherwise become atretic. The serious drawback of this form of therapy is that it is not easy to control the number of oocytes which are permitted to mature and ovulate—although careful monitoring of hormone levels can give warning of a multiple ovulation and enable multiple pregnancy to be avoided.

Failure to ovulate during lactation is associated with high levels of circulating prolactin (section 7.4). Many cases of infertility in women (and men) are likewise due to elevated prolactin levels, resulting in an impaired response to the gonads to FSH and LH, loss of the positive feedback response to oestrogen, and the loss of the episodic pattern of LH secretion. Treatment of infertile women with the recently-developed drug bromocriptine (a derivative of ergot, known from antiquity) lowers prolactin levels; the episodic pattern of LH reappears, the positive feedback

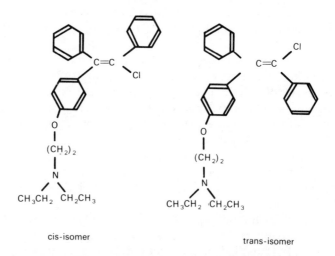

cis-isomer

trans-isomer

Figure 10.3 Structures of the two isomers of clomiphene.

response is reinstated, and ovulation and pregnancy may result. In one study, 80% of amenorrhoeic patients treated with bromocriptine resumed menstruation within 3 months; in another, 13 pregnancies were achieved in 12 patients following treatment. Bromocriptine appears to be effective even where infertility is not associated with high prolactin levels. Compared with treatments which involve the direct stimulation of ovulation, the use of bromocriptine to remove an inhibition should prove less hazardous; there is no reason to expect any multiple ovulations as, in effect, a normal menstrual cycle is being permitted to operate.

Male infertility is frequently associated with depressed testosterone levels and resultant abnormalities in sperm motility or concentration in the semen. A man will almost certainly be infertile if the sperm concentration is below 20×10^6 per ml semen, or if the concentration is higher but the sperm are relatively immotile. Sperm counts can in some cases be increased by hormone therapy. Androgen treatment is not often successful, possibly because androgens can inhibit pituitary gonadotrophin release, and thereby suppress spermatogenesis (section 10.3.2). An alternative is treatment with HCG (section 6.1.3) which acts like LH to stimulate endogenous testosterone output. In some cases, infertility is due, as in women, to abnormally elevated serum prolactin. Bromocriptine has proved effective in lowering prolactin levels and increasing fertility in men as in women.

In circumstances where normal insemination is not possible, or the semen is inadequate, artificial insemination can be an aid to conception. Semen can be donated either by the husband or (if he is irremediably sterile) by a suitable donor. Before artificial insemination (AI) could be widely used, techniques for the long-term storage of sperm were developed. Spermatozoa—and particularly structures such as the acrosome—are extremely delicate, so storage of fully functional sperm took some time to achieve. The most widely used methods involve diluting semen with a complex medium which includes glycerol and egg yolk. The composition of suitable dilutants was arrived at empirically, and it is still not entirely clear why these components are essential. After dilution, the semen may be cooled and frozen for storage in liquid nitrogen. In these conditions, it can remain capable of fertilization for a number of years.

The value of AI in alleviating infertility is shown by pregnancy rates of 50–70% achieved in one infertility clinic. The causes of infertility were various, including impotence, oligospermia (i.e. a sperm count of less than 20×10^6/ml) and impenetrable cervical mucus.

Long-term storage of human semen has also been suggested as an insurance against a subsequent change of mind by men about to undergo vasectomy, or to enable women to opt to bear children fathered (perhaps posthumously) by famous men. The low heritability of human abilities suggests that this might prove a disappointment.

In domesticated animals, artificial insemination is already very widely used. To take a single example, in 1975 AI accounted for more than 70% of all pigs inseminated in Britain. It is possible for one male, with desirable genetic attribrutes, to leave many more offspring, around the world, than he could ever have had the time to sire by more orthodox means.

10.7 Some future prospects

It is now possible to obtain a mature mammalian ovum, fertilize it *in vitro* and maintain the resulting embryo in culture for a few days. It can then be transferred back into a suitably prepared uterus for implantation and subsequent development. *In vitro* fertilization and culture of preimplantation embryos have made it possible to obtain much information about the metabolism of early development which would not otherwise have been available. In humans, the technique has been used practically in an attempt to bypass blocked oviducts in an infertile woman. Implantation occurred, but unfortunately in an oviduct and not in the uterus, and eventually the ectopic pregnancy had to be terminated. There seems to be no reason why fully successful implantation should not be achieved with human embryos cultured from eggs fertilized *in vitro*; a female baboon has recently given birth as a result of precisely this technique.

Mouse embryos can be cultured over the time at which implantation would normally have occurred *in vivo*. Extensive tissue differentiation occurs, but the embryos are always structurally disorganized. Rat embryos removed after implantation can be successfully cultured *in vitro* for much of their later development (figure 10.4). The uterine environment and the functions of the early yolk sac placenta can be duplicated to a large extent by the conditions of artificial culture. To date, it has been impossible to obtain growth of the allantoic placenta in culture. If this was achieved it would be possible to culture older embryos. If complete *in vitro* gestation proved possible for mouse or rat embryos, it should ultimately prove technically feasible with humans.

Culture of early embryos opens up the possibility of various forms of genetic manipulation. Ever since Aldous Huxley's *Brave New World* the idea of clones of genetically identical human beings has been a recurrent

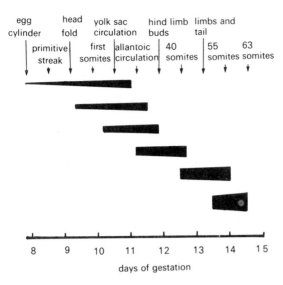

Figure 10.4 Periods of development which have been achieved *in vitro* with rat fetuses. (From New, D. A. T. (1973)): In Austin, C. R. (ed.): The Mammalian Fetus, *in vitro*. (Chapman and Hall).

nightmare, at least of the popular newspapers. It is possible to obtain such clones in amphibians. Nuclei are removed from cells of an early embryo and transplanted into enucleated eggs: the embryos which result are genetically identical. The human egg is so much smaller than that of an amphibian (100μm compared with over 1 mm in diameter) that it is unlikely that cloning could easily be achieved by this technique in the near future; quite apart from the improbability that women would readily volunteer to foster such embryos.

It is possible to transfer genetic material imto mammalian cells in culture, and to show that it is both expressed and replicated. One example is the successful transfer of a Chinese hamster gene for an enzyme (hypoxanthine ribosyl transferase) lacking in the recipient mouse fibroblasts. If the transfer of specific genes into human zygotes becomes possible, it might be an effective means of "gene therapy", that is, preventing the action of certain inherited defects or disorders of metabolism. An alternative approach might be to remove a proportion of embryonic cells, transform them appropriately, and return them to the original embryo, where they could now perform the previously deficient function on behalf of the entire embryo. A variant of this would be to permit a number of cells from a normal individual to colonize a genetically defective embryo and later perform the missing function.

Such an individual, in whom the defective genes would be masked would be a chimaera or composite of cells of two different genetic constitutions; he or she would be in the unusual situation of having a total of four parents. This sounds bizarre, but is in fact no different (except perhaps relatively) from the situation in which a person with a kidney disease or an immune deficiency receives a grafted kidney or a transfusion of competent lymphocytes. Both of these treatments are, if not exactly routine, at least regarded as acceptable.

Genetic transformation of the egg (hence of the entire individual) might lead to elimination of the defective gene; the infiltration of normal cells could mask its effects and permit its transmission to subsequent generations. From the eugenic point of view, the former, if it could be achieved, might therefore be preferable.

10.8 Ethics and the manipulation of human reproduction

In the preceding sections of this chapter, I have described the technical side of the present and potential methods of manipulating human reproduction. The social importance of reproduction has meant that numerous laws and ethical strictures have been applied to such methods. Insofar as legislation is merely an attempt to formalize a prevailing ethical view, I do not propose to consider it, but ethical attitudes to reproduction are themselves perhaps worth discussing further.

A number of religious creeds are interpreted as disapproving of any deliberate attempt to limit fertility. The Roman Catholic Church is among the most conspicuous, and has expressly condemned all forms of contraception except the singularly unreliable "rhythm" method, of limiting intercourse to the supposedly infertile period of the menstrual cycle. The reasoning behind this attitude is doubtless theologically impeccable, but it is not always easy for a lay public to follow, for instance, why a calendar should be more acceptable than a condom, or how the line of principle can be so finely drawn as to allow artificial steroid treatments which increase the chance of ovulation but exclude similar steroid treatments which suppress ovulation.

Disapproval of abortion is easier to understand. If an embryo is regarded as being, from the moment of fertilization, endowed with some property which makes it essentially a human being, then deliberate abortion involves the taking away of human life. If this is wrong after birth, it is surely so before.

The situation is, of course, not so simple. There are circumstances in which deliberate abortion might save the mother's life, and a failure to abort would result in the deaths of both mother and fetus. Most people would accept the prior claim to existence of the mother (although those who believe in original sin and the efficacy of baptism might disagree).

The decision is also not so simple when the fetus is known to be grossly deformed, and would inevitably have only a short and painful life. Many instances are arguable, but (for instance) an anencephalic fetus can never be normal and will certainly not survive long after birth. The same could be said of a number of severe and incurable genetic disabilities. Many people (including parents of such children) would accept abortion to be the preferable alternative.

In a separate category is the use of abortion as a means of birth control, with no eugenic purpose. On the one hand is the question of whether or not aborting a fetus is the same as killing a human being after birth; on the other, the welfare of the unwilling mother or of her unwanted baby, and the importance to society of reducing the birth rate.

It is unreasonable to impute human feelings to an embryo from the moment of fertilization. A zygote has no more human feelings than an amoeba. The development of a nervous system, or "quickening" (the first detectable fetal movements) might be more rational events from which to consider a fetus as an individual human being. An alternative might be the time at which a "postnatal" pattern of electrical activity can be detected in the brain; the converse of a criterion of death currently being canvassed in medical circles. Or perhaps the legal demarcation of 28 weeks is as valid as any; after this time, the fetus is potentially capable of an independent existence. However, thinking of humanity as a property which imbues the fetus at a definable stage of pregnancy is little more constructive than considering a soul to enter the oocyte with the sperm. To those willing at least to countenance the idea of abortion, the earlier the stage of pregnancy at which termination is carried out, the more acceptable.

Balanced against the rights imputed to a fetus are those of the pregnant woman. Not every unwanted pregnancy would result in an unwanted baby, but it is specious to use this as a *general* argument against abortion. The right of the mother to decide whether or not to have a child, and the effect of an unwanted birth on mother and baby, must be balanced against the arguable rights of the fetus. Recently, public opinion (and legislation) have moved towards attaching more importance to the former than to the latter.

There is the further consideration that women unable to obtain legal abortions may resort to illegal ones instead. Fewer than two maternal deaths result from every 100,000 abortions performed, in hospital conditions, within the first three months from conception and 12·2 per 100,000 during the second three months. Approximately 100 women die per 100,000 abortions performed out of hospitals by persons without medical training. Clearly most of these deaths are prevented by a liberal abortion policy.

It would, of course have been better for an unwilling mother to have used effective contraception or abstained from intercourse, but this is not an option once conception has occurred. Abortion is not the most desirable form of birth control; it is the only one which can be applied after conception.

In vitro fertilization and embryo culture, and genetic manipulations, also arouse strong emotions. As with the issues of contraception and abortion, opposition may be of two sorts. It can legitimately be maintained that such interference with "normal" processes is wrong in principle. This view can be accepted or rejected, but it is not open to argument. Alternatively, it may be asserted that such procedures are undesirable because the undesirable aspects outweigh the advantages.

The potential benefits are great; alleviating infertiliy, for example, or perhaps "gene therapy" to cure incapacitating genetic defects. The risks are that imperfect techniques might expose embryos to damage. Genetic cloning and complete gestation *in vitro* (if they ever become feasible) might also result in serious psychological problems for the children so produced. The decision to investigate human embryology and to apply the knowledge and technology so gained is not, therefore, a simple decision in principle, but must be made on the basis of an assessment of the relative risks and benefits in each case. Many people who accept *in vitro* fertilization as an aid to conception would recoil from the consequence of cloning of embryos: there are, however, advocates of cloning as a means of preserving particularly distinguished human genotypes which would otherwise die with their original owners.

The scientific study of human reproduction therefore presents choices not hitherto available. Although science may indicate the options and clarify the issues involved, it is not the function of science, or of scientists, to say what decisions should be made. Scientists are experts only on technical matters. When it comes to making the moral decisions, there are no experts. The responsibility for making such decisions must rest primarily with the individuals most concerned.

FURTHER READING

Chapter 1

Asdell, S. A. (1965), *Patterns of Mammalian Reproduction*, Constable.
Austin, C. R. and Short, R. V. (Eds.) *Reproduction in Mammals.*
    1. Germ cells and fertilization (1972).
    2. Embryonic and fetal development (1972).
    3. Hormones in reproduction (1972).
    4. Reproductive patterns (1972).
    5. Artificial control of reproduction (1972).
    6. The evolution of reproduction (1976).
                       (Cambridge Univ. Press).
Bentley, P. J. (1976), *Comparative Vertebrate Endocrinology*, Cambridge Univ. Press.
Cohen, J. (1977), *Reproduction*, Butterworth.
Conaway, C. H. (1971), "Ecological adaptation and mammalian reproduction" *Biology of Reproduction* **4**:239–47.
Coutinho, E. M. and Fuchs, F. (Eds.) (1974), *The Physiology and Genetics of Reproduction*, (2 volumes), Plenum.
Hogarth, P. J. (1976), *Viviparity* Edward Arnold/Institute of Biology.
Parkes, A. S. (Ed.) (1974/5) *Marshall's Physiology of Reproduction*, (3 volumes), Longman.
Tyndale-Biscoe, H. (1973), *The Life of Marsupials*, Arnold.
Williams, G. C. (1975), *Sex and Evolution*, Princeton Univ. Press.

Chapter 2

Beatty, R. A. (1970), "The genetics of the mammalian gamete", *Biological Reviews* **45**:73–119.
Duckett, J. G. and Racey, P. A. (Eds.) (1975), *The Biology of the Male Gamete*, Academic.
Johnson, A. D.; Gomes, W. R. and Vandemark, N. L. (1971), *The Testis*, (3 volumes), Academic.
Mann, T. (1964), *The Biochemistry of Semen and of the Male Reproductive Tract*, Wiley.
Open University (1974), "Gametes and the Physiology of Early Gestation", Course Unit 8 of Course S321 "The Physiology of Cells and Organisms" Open University Press.
Roosen-Runge, E. (1977), *The Process of Spermatogenesis in Animals*, Cambridge Univ. Press.
Waites, G. M. H. and Setchell, B. P. (1969), "Physiology of the testis, epididymis, and scrotum," *Advances in Reproductive Physiology* **4**:1–63.

Chapter 3

Austin, C. R. (1974), "Fertilization". pp. 48–75 in Lash, J. and 'Whittaker, J. R. (Eds.): *Concepts of Development*, Sinauer.
Blandau, R. J. (1968), "Gamete transport—comparative aspects". In Hafez, E. S. E. and Blandau, R. J. (Eds) *The Mammalian Oviduct*, Univ. Chicago Press.

174 FURTHER READING

Dewsbury, D. A. (1972), "Patterns of copulatory behaviour in male mammals", *Quarterly Review of Biology* 47:1–33.
Fox, C. A. and Fox, B. (1971), "A comparative study of coital physiology, with special reference to the sexual climax. Review", *Journal of Reproduction and Fertility* 24:319–36.
Longo, F. J. (1973), "Fertilization: a comparative structural review", *Biol. Reprod.* 9:149–215.
Metz, C. B. and Monroy, A. (Eds.) (1967), *Fertilization*, Academic.
Thibault, C. (1973), "Sperm storage and transport in vertebrates", *J. Reprod. Fertil. suppl.* 18:39–53.

Chapter 4

Arimura, A. and Findlay, A. (1971), "Hypothalamic map for the regulation of gonadotropin release", *Research in Reproduction* 3/1.
Catt, K. J. and Dufau, M. L. (1976), "Basic concepts of the mechanism of action of peptide hormones", *Biol. Reprod.* 14:1–15.
Goldberg, V. J. and Ramwell, P. W. (1975), "Role of prostaglandins in reproduction", *Physiological Reviews* 55:325–51.
Labhsetwar, A. P. (1974), "Prostaglandins and the reproductive cycle", *Federation Proceedings* 33:61–77.
Malkinson, A. M. (1975), *Hormone Action*, Chapman & Hall
Mossman, H. W. and Duke, K. C. (1973), *Comparative Morphology of the Mammalian Ovary*, Univ. Wisconsin Press.
Mossman, H. W. and Duke, K. C. (1974), *Comparative Morphology of the Mammalian Oviduct*, Univ. Wisconsin Press.
Nalbandov, A. V. (1970), "Comparative aspects of corpus luteum function", *Biol. Reprod.* 2:7–13.
O'Malley, B. W. and Schrader, W. T. (1976), "The receptors of steroid hormones", *Scientific American* 234/2:32–43.
Perry, J. S. (1971), *The Ovarian Cycle of Mammals*, Oliver and Boyd

Chapter 5

Balls, M. and Wild, A. E. (1975), *The Early Development of Mammals*, Cambridge Univ. Press.
Billington, W. D. (1971), "Biology of the trophoblast", *Adv. Reprod. Physiol.* 5:27–66.
Enders, A. C. (Ed.) (1963), *Delayed Implantation*, Univ. Chicago Press.
Finn, C. A. (1971), "The biology of decidual cells", *Adv. Reprod. Physiol.* 5:1–26.
Finn, C. A. and Porter, D. G. (1975), *The Uterus*, Elek.
McLaren, A. (1969), "Stimulus and response during early pregnancy in the mouse", *Nature* 221:739–41.
McLaren, A. (1976), *Mammalian Chimaeras*, Cambridge Univ. Press.
Moghissi, K. S. and Hafez, E. S. E. (Eds.) (1972), *Biology of Mammalian Fertilization and Implantation*, C. C. Thomas.
Nilsson, O. (1974), "The morphology of blastocyst implantation", *J. Reprod. Fertil.* 39:187–94.
Wimsatt, W. A. (1975), "Some comparative aspects of implantation", *Biol. Reprod.* 12:1–40.
Wolstenholme, G. E. W. and O'Connor, M. (Eds.) (1965), *Preimplantation Stages of Pregnancy*, Churchill.

Chapter 6

Eckstein, P. and Weir, B. J. (Eds.) (1976), "Fetal growth", *Journal of Reproduction and Fertility*. Symposium Report no. 8, *J. Reprod. Fertil.* 47:165–201.
Hytten, F. E. and Leitch, I. (1971), *The Physiology of Human Pregnancy*, Blackwell.
Klopfer, A. and Diczfalusy, E. (1969), *Foetus and Placenta*, Blackwell.

Open University (1974), "The physiology of the fetus and the new-born", Units 10 & 11 of Course S321 "Physiology of Cells and Organisms", Open University Press.
Ounsted, M. and Ounsted, C. (1973), *On Fetal Growth (its variations and their consequences)*, Heinemann.
Steven, D. H. (Ed.) (1975), *Comparative Placentation. Essays in structure and function*, Academic.
Wolstenholme, G. E. W. and O'Connor, M. (Eds.) (1969), *Foetal Autonomy* Churchill.

Chapter 7

Bonnar, J.: Franklin, M.: Nott, P. N. and McNeilly, A. S. (1975), "Effect of breast-feeding on pituitary-ovarian function after childbirth", *British Medical Journal* 1975/4:82–4.
Findlay, A. L. R. (1972), "The control of parturition", *Res. Reprod.* **4**/5.
Mepham, B. (1976), *The Secretion of Milk*, Arnold/Inst. of Biol.
Open University (1974), "Thermoregulation", Course Unit 12 of Course S321 "Physiology of Cells and Organisms", Open University Press.

Chapter 8

Fredga, K.; Gropp. A.; Winking, H. and Frank, F. (1976), "Fertile XX- and XY-type females in the wood lemming *Myopus schisticolor*", *Nature* **261**:225–7.
Grumbach, M. M.; Grave, G. D. and Mayer, F. E. (1974), *Control of the Onset of Puberty*, Wiley.
Herbert, J. (1976), "Hormonal basis of sex differences in rats, monkeys and humans", *New Scientist* **70**:284–6.
McDermott, A. (1975), *Cytogenetics of Man and other Animals*, Chapman & Hall.
Markert, C. L. and Papaconstantinou, J. (Eds.) (1975), *The Developmental Biology of Reproduction*, Academic.
Mittwoch, U. (1973), *Genetics of Sex Differentiation*, Academic.
Perry, J. S. (Ed.) (1969), "Intersexuality", *J. Reprod. Fertil.* Suppl. 7.

Chapter 9

Anderson, J. M. (1972), *Nature's Transplant*, Butterworth.
Anon, (1976), "Antibodies to spermatozoa", *Brit. med. J.* 1976/**2**:774–5.
Beer, A. E. and Billingham, R. E. (1974), "The embryo as a transplant", *Sci. Amer.* **230**/4:36–46.
Brambell, F. W. R. (1970), *The Transmission of Passive Immunity from Mother to Young*, North Holland.
Clarke, C. A. (1968), "The prevention of 'Rhesus' babies", *Sci. Amer.* **219**/5:46–52.
Edwards, R. G.; Howe, C. W. S. and Johnson, M. H. (1975), *Immunobiology of Trophoblast*, Cambridge Univ. Press.
Hemmings, W. A. (Ed.) (1976), *Maternofoetal Transmission of Immunoglobulins*, Cambridge Univ. Press.
Kirby, D. R. S. (1968), "Immunological aspects of pregnancy", *Adv. Reprod. Physiol.* **3**:33–80.
Scott, J. S. (Ed.) (1977), *Seminars in Perinatology* vol. 1 no. 2.
Scott, J. S. and Jones, W. R. (Eds.) (1976), *Immunology of Human Reproduction*, Grune & Stratton/Academic.

Chapter 10

Anon, (1975), "Hormone patterns in anorexia nervosa", *Brit. med. J.* 1975/**2**:52–3.
Briggs, M. H. and Briggs, M. (1976), *Biochemical Contraception*, Academic.

Bulmer, M. G. (1970), *The Biology of Twinning in Man*, Clarendon.

Edwards, R. G. (1974), "Fertilization of human eggs *in vitro*: morals, ethics and the law", *Quart. Rev. Biol.* **49**:3–26.

Edwards, R. G. and Sharpe, D. J. (1971), "Social values and research in human embryology", *Nature* **321**:87–91.

Ford, C. S. and Beach, F. A. (1952/1965), *Patterns of Sexual Behaviour*, Eyre & Spottiswoode

Fraser, F. C. and McKusick, V. A. (Eds.) (1970), *Congenital Malformations*, Excerpta Medica.

Hafez, E. S. E. and Evans, T. N. (Eds.) (1973), *Human Reproduction. Conception and Contraception*, Harper Row.

McKeown, T. (1976), *The Modern Rise of Population*, Arnold.

Noonan, J. T. (1966), *Contraception: a history of its treatment by the Catholic theologians*, Bellknap, Harvard Univ. Press.

Parkes, A. S. (1976), *Patterns of Sexuality and Reproduction*, Oxford Univ. Press.

Parkes, A. S.; Peel, J. and Thompson, B. (Eds.) (1971), "Biosocial aspects of human fertility", *Journal of Biosocial Science*, Suppl. 3.

Short, R. V. (1976), "The evolution of human reproduction", *Proc. Roy. Soc. B* **195**:3–24.

Short, R. V. and Baird, D. T. (Eds.) (1976), "A discussion on contraceptives of the future", *Proc. Roy. Soc. B* **195**:1–224.

General

Aiyer, M. S. and Fink, G. (1974), "The role of sex steroid hormones in modulating the responsiveness of the anterior pituitary gland to Luteinizing Hormone Releasing Factor in the female rat", *Journal of Endocrinology* **62**:553–72.

Amoroso, E. C. and Perry, J. S. (1975), "The existence during gestation of an immunological buffer zone at the interface between maternal and fetal tissues", *Phil. Trans. Roy. Soc. B* **271**:343–61.

Austin, C. R. and Weir, B. J. (Eds.) (1975), "Fertilization and cell fusion", *J. Reprod. Fertil.* Symposium Report No. 5; *J. Reprod. Fertil.* **44**:153–205.

Bedford, J. M. (1974), "Maturation of the fertilizing ability of mammalian spermatozoa in the male and female reproductive tract", *Biol. Reprod.* **11**:346–62.

Biggers, J. D. and Stern, S. (1973), "Metabolism of the preimplantation mammalian embryo", *Adv. Reprod. Physiol.* **6**:1–59.

Billington, W. D.; Kirby, D. R. S.; Owen, J. J. T.; Ritter, M. A.; Burtonshaw, M. D.; Evans, E. P.; Ford, C. E.; Gauld, I. K. and McLaren, A. (1969), "Placental barrier to maternal cells", *Nature* **224**:704–6.

Bindon, B. M.; Emmens, C. W. and Smith, M. S. R. (Eds.) (1973), "Research related to the control of fertility", *J. Reprod. Fertil.* suppl. 18.

Bogdanove, E. M.; Nolin, J. M. and Cambell, G. T. (1975), "Qualitative and quantitative gonad-pituitary feedback", *Recent Progress in Hormone Research* **32**:567–619.

Bogumil, R. J.; Ferin, M.; Rootenberg, J.; Speroff, L. and Vande Wiele, R. L. (1972), "Mathematical studies of the human menstrual cycle. I. Formulation of a mathematical model", *J. clin. Endocrinol. Metab.* **35**:126–43.

Bogumil, R. J.; Ferin, M. and Vande Wiele, R. L. (1972), "Mathematical studies of the human menstrual cycle. II. Simulation performance of the human menstrual cycle", *J. clin. Endocrinol. Metab.* **35**:144–56.

Carter, J. (1976), "The effect of progesterone, oestradiol and HCG on cell-mediated immunity in pregnant mice", *J. Reprod. Fertil.* **46**:211–6.

Castracane, V. D. and Rothchild, I. (1976), "Luteotrophic action of decidual tissue in the rat", *Biol. Reprod.* **15**:497–503.

Chard, T.; Hudson, C. N.; Edwards, C. R. W. and Boyd, N. R. H. (1971), "Release of oxytocin and vasopressin by the human foetus during labour", *Nature* **234**:352–4.

Clermont, Y. (1972), "Kinetics of spermatogenesis in mammals", *Physiol. Rev.* **52**:198–236.

Cohen, J. (1973), "Cross-overs, sperm redundancy and their close association", *Heredity* **31**:408–13.

Coutts, J. R. T. and Govan, A. D. T. (Eds.) (1975), "Functional morphology of the ovary", *J. Reprod. Fertil.* Symposium Report No. 7, *J. Reprod. Fertil.* **45**:557–619.

Denamur, R. (1974), "Luteotrophic factors in the sheep", *J. Reprod. Fertil.* **38**:251–60.

Dickmann, Z.: Gupta, J. S. and Dey, S. K. (1977), "Does 'blastocyst estrogen' initiate implantation?" *Science* **195**:687–8.

Diczfalusy, E. (Ed.) (1974), *Immunological Approaches to Fertility Control*, Karolinska Institutet, Stockholm.

Dobrowolski, W. and Hafez, E. S. E. (1971), "The uterus and control of ovarian function", *Acta Obstet. Gynecol. Scand.* Suppl. 12.

Edwards, R. G. (Ed.) (1969), *Immunology and Reproduction*, IPPF.

Edwards, R. G. (1970), "Immunology of conception and pregnancy", *Brit. Med. Bull.* **26**:72–8.

Erickson, R. P. (1977), "Androgen-modified expression compared with Y linkage of male specific antigen", *Nature* **264**:795–6.

de Fazio, S. R. and Ketchel, M. M. (1972), "The occurrence of seminal plasma antigens in the tissues of women", *J. Reprod. Fertil.* **30**:125–31.

Finn, R.; Hill, C. A. S.; Govan, A. J.; Ralfs, I. G.; Gurney, F. J. and Denye, V. (1972), "Immunological responses in pregnancy and survival of the fetal homograft", *Brit. Med. J.* 1972/**3**:150–2.

Ford, C. E. and Evans, E. P. (1977), "Cytogenetic observations on XX/XY chimaeras and a reassessment of the evidence for germ cell chimaerism in hetrosexual twin cattle and marmosets", *J. Reprod. Fertil.* **49**:25–33.

Goding, J. R. (1974), "The demonstration that $PGF_{2\alpha}$ is the uterine luteolysin in the ewe", *J. Reprod. Fertil.* **38**:261–71.

Gorski, R. A. (1971). "Sexual differentiation of the hypothalamus", in Mack, H. C. and Sherman, A. I. (Eds.), *The Neuroendocrinology of Human Reproduction*, C. C. Thomas.

Greep, R. O. (1973) (Ed.), *Female Reproductive System*, American Physiological Society Handbooks of Physiology Section 7, vol. II, parts 1 & 2.

Greep, R. O. and Porter, J. F. (Eds.) (1972), "Hypothalamic control of fertility", *J. Reprod. Fertil.* Suppl. 20.

Hamilton, D. W. and Greep, R. O. (Eds.) (1975), *Male Reproductive System*, Amer. Physiol. Soc. Handbooks of Physiology, Section 7, volume 5.

Harris, G. W. and Edwards, R. G. (Eds.) (1970), "A discussion on the determination of sex", *Phil. Trans. Roy. Soc.* **259**:1–206.

Hellström, I.; Hellström, K. E. and Allison, A. C. (1971), "Neonatally induced allograft tolerance may be mediated by serum-borne factors", *Nature* **230**:49–50.

Hetherington, C. M. and Humber, D. P. (1975), "The effects of active immunization on the decidual cell reaction and ectopic blastocyst development in mice", *J. Reprod. Fertil.* **43**:333–6.

Horton, E. W. and Poyser, N. L. (1976), "Uterine luteolytic hormone: a physiological role for prostaglandin $F_{2\alpha}$." *Physiol. Rev.* **56**:595–651.

James, W. H. (1976), "Timing of fertilization and sex ratio of offspring", *Ann. Hum. Biol.* **3**:549–56.

Johnson, D. C. (1972), "Sexual differentiation of gonadotropin patterns", *Amer. Zool.* **12**:193–205.

Katsh, S.; Aguirre, A. and Katsh, G. F. (1968), "Inactivation of sperm antigens by sera and tissues of the female reproductive tract", *Fertil Steril.* **19**:740–7.

Knobil, E. and Sawyer, W. H. (Eds.) (1974), *The Pituitary Gland and its Neuroendocrine Control*, Amer. Physiol. Soc. Handbooks of Physiology, Section 5, volume 4.

Krarup, T.; Pederson, T. and Faber, M. (1969), "Regulation of oocyte growth in the mouse ovary", *Nature* **224**:187–8.

M

Lau, H. L. and Linkins, S. E. (1976), "Alpha-fetoprotein", *Amer. J. Obstet. Gynecol.* **124**:533–49.

Levine, N. and Marsh, D. J. (1975), "Micropuncture study of the fluid composition of 'Sertoli-cell-only' seminiferous tubules in rats", *J. Reprod. Fertil.* **43**:547–9.

Liggins, G. C.; Forster, C. S.; Grieves, S. A. and Schwartz, A. L. (1977), "Control of parturition in man", *Biol. Reprod.* **16**:39–56.

MacLeod, J. (1970), "The significance of deviations in human sperm morphology", *Adv. exp. Med. Biol.* **10**:481–94.

Murgita, R. A. and Tomasi, T. B. (1975), "Suppression of the immune response by α-fetoprotein", *J. exp. Med.* **141**:269–86.

Muggleton-Harris, A. L. and Johnson, M. H. (1976), "The nature and distribution of serologically detectable alloantigens on the preimplantation mouse embryo", *J. Embryol. Exp. Morphol.* **35**:59–72.

Odeblad, E. (1968), "The functional structure of human cervical mucus", *Acta Obstet. Gynecol. Scand.* **47** suppl. 1:57–79.

Ohno, S.; Christian, L. C.; Wachtel, S. S. and Koo, G. C. (1976), "Hormone-like role of H-Y antigen in bovine freemartin gonad", *Nature* **261**:597–9.

O'Malley, B. W. and Means, A. R. (1973), *Receptors for Reproductive Hormones*, Plenum.

Parker, C. R. and Mahesh, V. B. (1976), "Hormonal events surrounding the natural onset of puberty in female rats", *Biol. Reprod.* **14**:347–53.

Perry, J. S. (Ed.) (1971), "Spermatogenesis and sperm maturation", *J. Reprod. Fertil.* suppl. 13.

Perry, J. S. (1971), (Ed.) "Experiments on mammalian eggs and embryos", *J. Reprod. Fertil.* Suppl. 14.

Perry, J. S. (Ed.) (1972), "Control of parturition", *J. Reprod. Fertil.* suppl. 16.

Perry, J. S. (Ed.) (1974), "Secretions of the male and female reproductive tracts", *J. Reprod. Fertil.*, Symposium Report No. 1, *J. Reprod. Fertil.* **37**:163–250.

Perry, J. S. (Ed.) (1974), "Implantation", *J. Reprod. Fertil.*, Symposium, Report No. 3, *J. Reprod. Fertil.* **39**:173–249.

Perry, J. S. (Ed.) (1974), "Prolactin", *J. Reprod. Fertil.*, Symposium Report No. 4, *J. Reprod. Fertil.* **39**:437–99.

Pierrepoint, C. G. and Weir, B. J. (Eds.) (1975), "Endocrinology of the male genital tract", *J. Reprod. Fertil.* Symposium Report No. 6, *J. Reprod. Fertil.* **44**:335–409.

Restall, B. J. (1967), "The biochemical and physiological relationships between the gametes and the female reproductive tract", *Adv. Reprod. Physiol.* **2**:181–212.

Richards, J. S. and Midgley, A. R. (1976), "Protein hormone action: a key to understanding ovarian follicle and luteal cell development", *Biol. Reprod.* **14**:82–94.

Roberts, C. J. and Lowe, C. R. (1975), "Where have all the conceptions gone?" *The Lancet* 1975:498.

Rodger, J. C. (1975), "Seminal plasma, an unnecessary evil?" *Theriogenol.* **3**:237–47.

Rondell, P. (1970), "Follicular processes in ovulation", *Fedn. Procs.* **29**:1875–9.

Roosen-Runge, E. C. (1973), "Germinal cell loss in normal metazoan spermatogenesis", *J. Reprod. Fertil.* **35**:339–48.

Salisbury, G. W.; Hart, R. G. and Lodge, J. R. (1976), "The fertile life of spermatozoa", *Perspectives Biol. Med.* **19**:213–230.

Schally, A. V.; Arimura, A. and Kastin, A. J. (1973), "Hypothalamic regulatory hormones", *Science* **179**:341–50.

Schlafke, S. and Enders, A. C. (1975), "Cellular basis of interaction between trophoblast and uterus during implantation", *Biol Reprod.* **12**:41–65.

Setchell, B. P. and Main, S. J. (1974), "Bibliography (with review) on Inhibin", *Bibliog. Reprod.* **24**/3:245–52 and 361–7.

Shapiro, B. H.; Goldmann, A. S.; Steinbeck, H. F. and Neumann, F. (1976), "Is feminine determination of the brain hormonally determined?" *Experientia* **32**:650–1.

Steinberger, E. (1971), "Hormonal control of mammalian spermatogenesis", *Physiol. Rev.* **51**:1–22.

Steinberger, A. and Steinberger, E. (1976), "Secretion of an FSH-inhibiting factor by cultured Sertoli cells", *Endocrinol.* **99**:918–21.

Thorner, M. O.; Besser, G. M.; Jones, A.; Dacie, J. and Jones, A. E. (1975), "Bromocriptine treatment of female infertility: report of 13 pregnancies", *Brit. med. J.* 1975/**4**:694–7.

Wales, R. G. (1975), "Maturation of the mammalian embryo: biochemical aspects", *Biol. Reprod.* **12**:66–81.

Wallis, M. (1975), "The molecular evolution of pituitary hormones", *Biol. Rev.* **50**:35–98.

Weiss, G.; O'Byrne, E. M. and Steinetz, B. G. (1976), "Relaxin: a product of the human corpus luteum of pregnancy", *Science,* **194**:948–9.

Wood, W. G.; Pearce, K.; Clegg, J. B.; Weatherall, D. J.; Robinson, G. S.; Thorburn, G. D. and Dawes, G. S. (1976), "Switch from foetal to adult haemoglobin synthesis in normal and hypophysectomised sheep", *Nature* **264**:799–801.

Yochim, J. M. (1975), "Development of the progestational uterus: metabolic aspects", *Biol. Reprod.* **12**:106–33.

Zolovick, A. and Labhsetwar, A. P. (1973), "Evidence for the theory of dual hypothalamic control of ovulation", *Nature* **245**:158–9.

Index

abnormal development, detection of 163–4, 171
ABO blood groups: *see* blood groups
abortion 162–4, 170–2
ABP: *see* Androgen Binding Protein
accessory organs 24–7: *see* seminal vesicles, prostate, Cowper's glands, Littré's glands
acriflavine 19
acrosin 52–3, 161
acrosome 11–13, 49–53, 138
  enzymes of 49–52
  reaction 49–52
acrozonase: *see* acrosin
ACTH: *see* adrenocorticotrophic hormone
actinomycin 86
activation of spermatozoa, embryos: q.v.
activity during female cycle 74, 124
adaptation of newborn: *see* neonate
adenohypophysis (anterior pituitary): *see* pituitary
adenosine monophosphate, cyclic (cAMP) 28, 63, 144
adenosine triphosphate (ATP) 28, 104
adhesion of embryo 78, 80, 84
adjuvant 145
adrenals, adrenal cortex 75, 92
  fetal 94–5, 105–6, 108, 109, 128
  hormones of 75, 106, 113, 128: *see* corticosteroids
adrenocorticotrophic hormone (ACTH) 105–6, 108, 109, 149
AFP: *see* α–fetoprotein
Agnatha 3
AI: *see* artificial insemination
alkaline phosphatase 83
alkaloid 105
allantois 96, 118, 168–9
allergy 134, 143
allograft 135, 139–40
  fetus as an 4, 148–52
amenorrhoea 156
amino acids 4–7, 20–1, 23, 30, 46, 88, 102, 114, 133
amniocentesis 163–4
amnion, amniotic fluid 162–3
Amphibia 4, 169

amygdala 131
anamnestic response 133
androgen 27–35, 60, 68, 75, 109, 119, 121, 123, 125–6, 128, 161–2, 167: *see also* 5α–androstanediol, androstenedione, testosterone
Androgen Binding Protein (ABP) 29–30
androgenization 125–6
5α–androstanediol 29
androstenedione 62, 75, 106
anencephaly 105, 108, 163–4, 171
anoestrus 57, 75
anorexia nervosa 131
antiandrogen 32: *see* cyproterone acetate
antibody 1, 27, 52, 70, 113, 133–9, 141–5, 147–54, 162: *see also* immunoglobulins, immunity
  uptake from gut 153–4
antigen 132–7, 138–49: *see also* histocompatibility antigen, blood groups
antigenicity 132
  of embryo, fetus 148–52
  of semen 137–42, 144–5, 148
antigonadotrophin 35
antioestrogen 86, 165–6
antiserum 92
arginine vasotocin 5, 7, 35
armadillo 89
Artibeus 89
artificial insemination (AI) 74, 165, 167–8
ATP: *see* adenosine triphosphate
atresia 60–2
autoimmunity 145, 148
autosome 117, 120
azoospermia 141, 145, 161–2

baboon 40–41, 168
badger 88–9
baldness 28
barbiturates 70, 126: *see* phenobarbitol
Barr body 117
basal body temperature 68, 165
bat 21, 26, 47–8, 88–9, 145
BAT: *see* brown adipose tissue
B–cells (lymphocytes) 136, 152

184

168–9: *see also* DNA, histocompatibility antigens, sex determination
genotype, expression of spermatozoal 16–7, 139, 140
germ cell 9, 147: *see* oocyte, oogonia, spermatids, spermatocytes, spermatogonia, spermatozoa
  cell death 15–6, 58–60, 62
  cells, primordial 58, 118–121, 127
gestation: *see* pregnancy
GH: *see* Growth Hormone
β–globulin 141
γ–globulin: *see* immunoglobulins
glucocorticoid: *see* corticosteroids
glucose 20–1, 23, 45–6, 78–9, 85, 88, 101–2. 111. 114. 115
glycerylphosphorylcholine (GPC) 22, 46
glycogen 46, 48, 83
glycoprotein 1, 4, 6, 22, 38–9, 50, 93, 138, 143
  hormones: see LH, FSH, prolactin, HCG
goat 107, 112, 120, 153
gonad 4, 6: *see* ovary, testis
  determination of sex of 118–121
  primordia 118–121, 127
gonadal hormones: *see* steroids, relaxin
gonadotrophins 86, 158, 161–2, 166: *see* pituitary anterior hormones, placenta gonadotrophins, FSH, LH, prolactin
GPC: *see* glycerylphosphorylcholine
Graafian follicle 60–1, 63, 67
graft rejection: *see* immune response
granulosa cells 48, 60–2, 63: *see* corpus luteum
grey seal 88
grizzly bear 36
Growth Hormone, pituitary 4, 96, 113
guinea pig 44, 47, 51, 94, 101

H–2 antigen 139
Halichoerus 88
hamster 50, 58, 169
H–antigen: *see* histocompatibility antigen
HCG: *see* human chorionic gonadotrophin
HCS: human chorionic somatomammotrophin: synonym for HPL (q.v.)
heat 55, 75: *see* oestrus
hedgehog 153
Herpestes 117
heterogametic sex 116–7, 140
hippopotamus 5
histocompatibility antigen 138–141, 143–4, 148–50, 152: *see* H–2, HL–A, H–Y
HL–A 140
hormone 1–7: *see* glycoprotein, gonado-

trophins, corpus luteum, placenta, pituitary, oligopeptides, steroids
hormone antagonist: *see* antiandrogen, antigonadotrophin, antioestrogen
  receptors 28–9, 34, 57–8, 63–64, 70–1, 119, 127: *see* Androgen Binding Protein, α–fetoprotein
horse 21, 47, 60, 62, 80, 96, 99–100, 153
HPL: *see* human placental lactogen
human 2, 4, 13, 15, 21, 24–6, 34, 36–9, 42, 44–5, 47, 50, 54–5, 58–9, 62, 67–8, 72–6, 90, 93–7, 99–102, 105, 107, 110–1, 113, 115, 116–7, 121, 125, 128–31, 138–40, 142–154, 155–72
  chorionic gonadotrophin (HCG) 4, 6, 93–4, 96, 149, 161, 167
  chorionic somatomammotrophin (HCS) synonym for human placental lactogen (q.v.)
  placental lactogen (HPL) 4, 94, 96–7, 102, 149
H–Y antigen 119–120, 140–1
hyaluronic acid 52
hyaluronidase 52
hydrogen peroxide 25
hypophysectomy (removal of pituitary) 27–8, 30, 64, 92, 98, 105–6
hypophysiotrophic area of hypothalamus 70, 72, 126: *see also* Median Eminence
hypophysis: synonym for pituitary (q.v.)
hypothalamic releasing factors 1, 6, 109: *see* corticotrophin releasing factor, LH/FSH releasing factor, prolactin inhibitory factor
hypothalamus 1, 6, 30–1, 67, 70–1, 75, 108–9, 158, 161: *see* hypophysiotrophic area, Median Eminence, preoptic area
  sexual differentiation of 126–8
hysterectomy 65

ICSH: interstitial cell stimulating hormone: synonym for luteinizing hormone (q.v.)
immune enhancement 4, 152
immune response 119, 132–7
  modulation of 148–9, 152
  of female against semen 15, 26, 48, 132, 137, 141, 142–8, 152
  of female against embryo or fetus 4, 132, 148–52
  of male against spermatozoa 145, 147, 162
immune tolerance 135, 144
immunity, passive acquisition of 113, 153–4
immunofluorescence 136, 138–9, 142
immunoglobulins (Ig) 133–4, 154: IgA 134, 142–3, 145–6, IgD 134: IgE 134, 142–3: IgG 133–4, 142–6, 153–4: IgM 134, 142–4, 153–4